T0234602

Small-scale Map Projection Design

Research Monographs in
Geographic Information Systems
Edited by Peter Fisher and Jonathan Raper

Small-scale Map Projection Design

Frank Canters

CRC Press
Taylor & Francis Group
Boca Raton London New York

CRC Press is an imprint of the
Taylor & Francis Group, an **informa** business

A TAYLOR & FRANCIS BOOK

First published 2002 by Taylor & Francis

Published 2020 by CRC Press
Taylor & Francis Group
6000 Broken Sound Parkway NW, Suite 300
Boca Raton, FL 33487-2742

First issued in paperback 2020

© 2002 by Frank Canters
CRC Press is an imprint of Taylor & Francis Group, an Informa business

No claim to original U.S. Government works

ISBN 13: 978-0-367-57864-0 (pbk)
ISBN 13: 978-0-415-25018-4 (hbk)

Visit the Taylor & Francis Web site at
http://www.taylorandfrancis.com

and the CRC Press Web site at
http://www.crcpress.com

Every effort has been made to ensure that the advice and information in this book is true and accurate at the time of going to press. However neither the publisher nor the authors can accept any legal responsibility or liability for any errors or omissions that may be made. In the case of drug administration, any medical procedure or the use of technical equipment mentioned within this book, you are strongly advised to consult the manufacturer's guidelines.

Publisher's Note
This book has been prepared from camera-ready copy provided by the author.

British Library Cataloguing in Publication Data
A catalogue record for this book is available from the British Library

Library of Congress Cataloging in Publication Data
Canters, Frank.
    Small scale map projection design/ Frank Canters.
      p. cm.
    Includes bibliographical references.
1.Map projection – Data processing. 2.Geographic information systems.
I.Title. II. Series.

GA110.C335 2001   2001053516
526'.8--dc21

Contents

Series introduction

WELCOME

The *Research Monographs in Geographical Information Systems* series provides a publication outlet for research of the highest quality in GIS, that is longer than would normally be acceptable for publication in a journal. The series includes single- and multiple-author research monographs, often based upon PhD theses and the like, and special collections of thematic papers.

THE NEED

We believe that there is a need, from the point of view of both readers (research and practitioners) and authors, for longer treatments of subjects related to GIS than are widely available currently. We feel that the value of much research is actually devalued by being broken up into separate articles for publication in journals. At the same time, we realise that many career decisions are based on publication records, and that peer review plays an important part in that process. Therefore a named editorial board supports the series, and advice is sought from them on all submissions.

Successful submissions will focus on a single theme of interest to the GIS community, and treat it in depth, giving full proofs, methodological procedures or code where appropriate to help the reader appreciate the utility of the work in the Monograph. No area of interest in GIS is excluded, although material should demonstrably advance thinking and understanding in spatial information science. Theoretical, technical and application-oriented approaches are all welcomed.

THE MEDIUM

In the first instance the majority of Monographs will be in the form of a traditional textbook, but, in a changing world of publishing, we actively encourage publication on CD-ROM, the placing of supporting material on web sites, or publication of programs and of data. No form of dissemination is discounted, and the prospective authors are invited to suggest whatever form of publication and support material they think is appropriate.

THE EDITORIAL BOARD

The Monograph series is supported by an editorial board. Every monograph proposal is sent to all members of the board which includes Ralf Bill, António Câmera, Joseph Ferreira, Pip Forer, Andrew Frank, Gail Kucera, Peter van Oostrom, and Enrico Puppo. These people have been invited for their experience in the field, of monograph writing, and for their geographic and subject diversity. Members may also be involved later in the process with particular monographs.

FUTURE SUBMISSIONS

Anyone who is interested in preparing a Research Monograph should contact either of therse editors. Advice on how to proceed will be available from them, and is treated on a case by case basis.

For now we hope that you find this, the tenth in the series, a worthwhile addition to your GIS bookshelf, and that you may be inspired to submit a proposal too.

Editors:

Professor Peter Fisher
Department of Geography
University of Leicester
Leicester
LE1 7RH
UK
Phone: +44 (0) 116 252 3839
Fax: +44 (0) 116 252 3854
Email: pff1@le.ac.uk

Professor Jonathan Raper
Schools of Informatics
City University
Northampton Square
London
UK
Phone: +44 (0) 20 7477 8000
Fax: +44 (0) 20 7477 8587
Email: raper@soi.city.ac.uk

Preface

I cannot recall exactly when my fascination for maps began. As a child, still very young, I took great pleasure in browsing through my grandfather's collection of old pre-war atlases, which he kept in one of his drawers. For hours I would leaf through the dusty, fading pages of these fascinating remnants of the past, bearing evidence of countries and places with exotic names I had never heard of, and revealing intriguing details about the appearance and habits of distant people and cultures. This early fascination for maps carried on in the years that followed, and by the time I finished high school it became clear to me that I was destined to do something that, in one way or the other, had to do with maps. So it is probably no surprise that I finally ended up studying geography.

After my graduation, I started a PhD project on small-scale map projection design under the supervision of Hugo Decleir. In the first years of the project, research mainly focused on the quantification of map projection distortion on small-scale maps, which led to the publication of *The World in Perspective*, a comprehensive survey of the distortion characteristics of well known and less well known projections for world maps. From this research, it became clear to me that traditional approaches to map projection design, which are exclusively based on the study of distortion characteristics at the infinitesimal scale, are of limited use for the development of small-scale map projections, and that a more versatile approach is required. Next to map projection distortion at the infinitesimal scale, one should take account of the impact of local scale variation on the distortion of areas of finite size. At the same time, one should also pay attention to non-quantifiable features the map projection should have, mostly in relation to the purpose and/or the use of the map. These ideas are essential to the work that is reported in this monograph.

A large part of this book is taken from my PhD thesis, but new material has been added to reflect the present state of the art in the field, and to provide more background on some of the issues that are discussed. While most reference works on map projection focus on purely mathematical aspects, this monograph puts theoretical research next to map projection use, and aims to bring the academic and the map maker closer together. It is my hope, therefore, that the book will appeal to both researchers and practitioners engaged in the design of small-scale maps. It may also be used as a supplementary text for advanced courses concerned with map projection, analytical cartography, or map design.

It took me a long time to finish the work that is reported in this monograph. After a few years of research, I gradually became more involved in GIS- and RS-related projects and teaching activities, making it difficult to find the time to continue my work on map projections. Had there not been my many friends and colleagues who encouraged and supported me, I would probably never have succeeded in finishing this study. I am grateful to all of them. They are simply too numerous to mention. There are, however, two people to whom I would like to express my special thanks. First of all, I would like to thank my mentor Hugo Decleir, who introduced me to the field of map projections. Especially in the first years of my PhD work, his help and advice were of great value to me. I also thank my colleague William De Genst, with whom I shared part of the map projection research, for the many discussions, comments and suggestions, and for enduring my moods in all the years that passed since we started sharing the office.

My thanks also go to all map projection scientists and enthusiasts who over the years took an interest in my work, and supplied me with useful information. I am grateful to each and every one of them. In particular, I would like to thank Christoph Brandenberger for his critical remarks on my PhD work, Waldo Tobler and Mark Monmonier for constructive comments on the content of the manuscript for this book, and especially the late John Snyder, who was without any doubt my greatest source of inspiration, and the one who taught me the most about map projections, even though I only got the chance to meet him on a couple of occasions.

This research would not have been possible without the financial support of the Flanders' Fund for Scientific Research (F.W.O.-Vlaanderen), who offered me a research grant for a period of four years. I would like to thank the then members of the Scientific Commission on Earth Sciences who positively advised the Fund with regard to my research proposal, and gave me the opportunity to start my career as a researcher.

I thank my dear friend and Adobe™ wizard Geert for preparing many of the figures for the final version of the manuscript. Finally, I thank Dimi, for doing the layout for this book, but mostly for comforting and supporting me, and reminding me of what is most important in life.

FRANK CANTERS
Brussels

Introduction

From the early days of map making to the present time, the challenge of representing the round Earth or part of it on a flat piece of paper without introducing excessive distortion has attracted the attention of many geographers, physicists, astronomers, mathematicians, and map makers. Putting aside the Medieval period, when most representations of the Earth's surface were influenced by religious ideology and loaded with symbolic meaning (think of the famous East-oriented T-O maps that depict the Earth as a round disk, subdivided in three continents by a T-shaped sea, and surrounded by an O-shaped ocean), the accurate representation of the Earth as it is known at a particular time has always been an important objective of contemporary map making. This does not imply that the subject of map projection has received an equal amount of interest from the ancient Greek period, when Claudius Ptolemy wrote his famous *Geography*, all the way up to the twentieth century. Major breakthroughs in the history of map projections were prompted by various external factors such as, for example, the increasing geographical knowledge during the Renaissance, which led to modifications of earlier map projections and the development of a whole series of new map projections suitable for displaying the entire globe, or the development of the calculus (first applied to map projections by Johann Heinrich Lambert in the late eighteenth century), which gave the cartographer or mathematician the necessary tools for the development of new map projections that fulfil certain general conditions, the most important of these being the preservation of angles and area.

The introduction of modern computers marks a new era in the evolution of map projection science. Having been dominated for almost two centuries by the formulation of analytical solutions to increasingly complex map projection problems, the computer cleared the way for a numerical treatment of map projection. Numerical approaches have been used to solve various practical problems related to map projection, including the efficient transfer of data from one map projection to another (Doytsher and Shmutter, 1981; Wu and Yang, 1981; Snyder, 1985; Kaltsikis, 1989; Canters, 1992), and the automated identification of map projection type and/or map projection parameters for maps for which this information is not known (Snyder, 1985). The computer has also been used to calculate and compare distortion on various map projections (Tobler, 1964; Francula, 1971; Peters, 1975; Canters and Decleir, 1989), and to automatically determine the value of the parameters of standard map projections so that overall scale error is reduced (Snyder, 1978a; Grafarend and Niermann, 1984; Snyder, 1985). One of the most stimulating outcomes of computer-assisted map projection research so far has been the development of conformal map projections with distortion patterns that are closely adapted to the shape of the area to be mapped, guaranteeing low overall scale error within the approximate boundaries of the area (Reilly, 1973; Stirling, 1974; Lee, 1974; Snyder, 1984a, 1985, 1986; González-López, 1995; Nestorov, 1997). Although the theoretical foundations for the development of these projections and the first simple applications of the principle date from the pre-computer age (Laborde, 1928; Driencourt and Laborde, 1932; Miller, 1953), a more complicated use of the technique requires extensive numerical processing, practical only in a digital setting.

The digital revolution has also had a large impact on the everyday use of map projections. As computer-based mapping tools make it possible for an ever growing group of people without any formal cartographic background to make their own maps, it also gives them the freedom to become more creative with map projections. Complicated map projections that are difficult to draw manually, and for that reason were seldom used in the past, can now be plotted quickly, using map projection software libraries or map projection tools that come with standard GIS software. Projection parameters can easily be changed, making it possible to experiment with alternative views of the same geographical area with very little effort. Although this may lead to a greater awareness of the pros and cons of using a particular map projection for a given purpose, it also creates much opportunity for misuse. At present, no commercially available map projection software offers the user some basic guidance in choosing a proper map projection for a particular mapping task. This is very unfortunate, especially in small-scale mapping, where the map maker has a large number of potentially useful map projections to choose from. When constructing a small-scale map, the choice of projection may determine to a large extent if the data are portrayed adequately in relation to the purpose of the map and, accordingly, if the map fulfils its role as a communicative device or not (see for instance American Cartographic Association, 1991). Examples of bad map design are quite common, in print journalism as well as in the electronic media (Gilmartin, 1985; Monmonier, 1989).

Offering the map maker some assistance in selecting a proper map projection might take many forms, from the provision of relatively simple tools for the display of map projection distortion and/or the optimisation of projection parameters, to the development of expert systems that guide the map maker through the entire selection process, by asking him or her a number of appropriate questions related to map function and map use. Several attempts have been made to fully automate the map projection selection process, yet none of these has become operational (Jankowski and Nyerges, 1989; Smith and Snyder, 1989; Kessler, 1991). One of the major difficulties in small-scale map projection selection is the great number of established map projections one can choose from. While some map projections have very distinct properties, and are without any doubt the best choice for particular mapping tasks, others have very similar characteristics, making it difficult to decide which map projection is best for a given purpose. Having to deal with such a large set of map projections substantially complicates the selection process. On the other hand, even a large number of map projections only represents a small subset of all theoretically possible mappings, so any selection process that is based on such a set will, by definition, produce sub-optimal solutions.

The main objective of this monograph is to show how numerical techniques can be used to assist the map maker in choosing or developing a proper projection for a small-scale map, taking into account the purpose of the mapping. One of the most important features of the approach that is presented in this book is its emphasis on map projection optimisation. Rather than starting out with a fixed set of map projections and trying to choose the most appropriate one, map projection selection is presented as a dynamic, computer-assisted design process that is controlled by the map maker. In the process, which starts with a "trial" projection that has all the properties that are required, map projection features and graticule geometry are iteratively changed and adapted to the area to be mapped. This ultimately results in a unique "tailor-made" projection – the term was first suggested by Robinson in 1974 – for each particular application. By proposing techniques for the optimisation of map projections, building on recent advances in map projection research, and showing how these techniques can be used in the process of map projection selection, this study also tries to narrow the gap between theoretical research on map projections, which is seldom promoted beyond scientific publication, and map projection use, which far too often relies on the application of simple rules-of-thumb.

The book consists of six chapters. The first chapter summarises the most important elements of map projection theory, and reviews different approaches to map projection classification. This provides the reader with the necessary background information for further reading.

The second chapter deals with the definition and application of various measures of map projection distortion, derived from local distortion theory. Distortion is analysed in detail for a large number of standard map projections used for world maps. This permits us to draw some conclusions about the impact of map projection geometry on the distortion pattern of a projection, information that may be helpful in making the right decisions about graticule geometry when selecting a projection. The results of the analysis also allow us to point out some of the limitations of map projection evaluation based on local measures of distortion.

Chapter 3 discusses the problems associated with the measurement of map projection distortion at the finite scale. Different approaches for characterising finite map projection distortion are reviewed and criticised, and a new, improved measure of finite scale distortion is proposed. Also investigated is how this measure relates to both qualities we expect from a small-scale map, i.e. the maintenance of true proportions and the proper representation of shape.

Chapter 4 reviews different methods for the reduction of map projection distortion, including modification of map projection parameters, derivation of optimal projections from generalised map projection equations, and transformation of map projection coordinates using polynomials. Special attention is paid to the definition or preservation of geometric and other properties when modifying or transforming an existing map projection. Transformation of map projections without loss of their original properties is of particular importance when we are thinking of incorporating map projection optimisation into the process of map projection selection, which is the ultimate goal of this study.

In chapter 5 most of the transformations methods discussed in the previous chapter are applied to produce new low-error map projections satisfying different sets of conditions. In all the examples shown the finite distortion measure, introduced in chapter 3, is used as a criterion for optimisation. Examples include a series of new map projections for the mapping of the world, optimised equal-area projections for Europe and Africa, and new low-error, equal-area projections for the European Union.

Chapter 6 starts with a comprehensive review of different approaches to map projection selection proposed so far. Next, the most important features of a map projection and their use are summarised. Based on this set of features, a possible strategy for small-scale map projection selection is presented. The proposed strategy links rule-based selection (to define the map projection type) with numerical optimisation (to reduce overall distortion), making use of the techniques described in the previous chapters.

It is hoped that this work will contribute to greater awareness of the decisive role of the map projection in small-scale map design, and to a more creative and well-considered use of map projections in general. It may also inspire future developers of map projection tools to include methods for map projection evaluation and optimisation in their products, and offer the map maker some assistance in choosing a projection that fits the purpose of the map.

CHAPTER ONE

Fundamentals
of map projection theory

The projection of a curved surface on a plane map always introduces distortion. As a result, map scale varies from one location to another and is generally different in every direction. When trying to represent the Earth on a plane the primary concern is to choose a map projection on which distortion is a minimum. Accomplishing this necessitates an understanding of how distortion takes place and how it is distributed over the entire map area. The mathematical theory of map projection provides us with the necessary tools to study the distortion characteristics of any map projection. The fundamentals of this theory were laid down in the late 18th century by Johann Heinrich Lambert, and further elaborated in the 19th century, mostly by Carl Friedrich Gauss and Nicolas Auguste Tissot.

In this chapter, the most important elements of map projection theory will be briefly summarised. This will provide the necessary background for further reading. Those who are interested in full mathematical derivations are referred to Canters and Decleir (1989). Since this study is only concerned with small-scale map projection, it is assumed that the Earth is approximated by a spherical model with known radius. For a more general, ellipsoidal treatment of map projection theory the reader is referred to Hoschek (1969), Richardus and Adler (1972), and Bugayevskiy and Snyder (1995).

Along with the mathematical theory of map projection, this chapter also pays attention to the classification of map projections. Classification facilitates a systematic treatment of the subject of map projection and often provides a useful context for the development of strategies for map projection selection. While the *Bibliography of Map Projections* lists twenty-three references on the subject (Snyder and Steward, 1988), it is not an objective of this study to offer a complete review of all attempts to classification. Rather a distinction will be made between different approaches that have been taken, with examples illustrating the merits and shortcomings of each of them. Only those classifications that are most often referred to will be treated in more detail.

1.1   DEFINITION OF A MAP PROJECTION

The whole process of representing the Earth on a map can be schematised in three stages (figure 1.1):

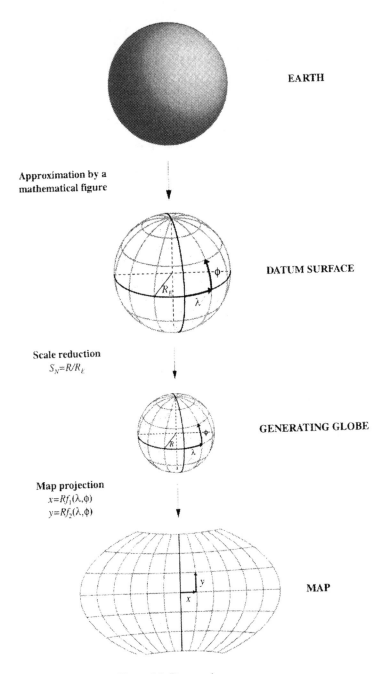

EARTH

Approximation by a
mathematical figure

DATUM SURFACE

Scale reduction
$S_N = R/R_E$

GENERATING GLOBE

Map projection
$x = Rf_1(\lambda, \phi)$
$y = Rf_2(\lambda, \phi)$

MAP

Figure 1.1 The mapping process.

1. First the size and shape of the Earth is approximated by a mathematical figure, a datum surface, for small-scale mapping a sphere with radius $R_E = 6371$ km.

2. Since a map is a small-scale representation of the Earth a scale reduction must take place which transforms the spherical model mentioned above into a smaller sphere that is called the *generating globe*. The *principal scale* (also called *nominal scale*) of the map S_N is then defined as the ratio of the radius of the generating globe R to the radius of the spherical model R_E :

$$S_N = R/R_E \tag{1.1}$$

3. Finally, the map projection converts the generating globe into a map. The number of ways of accomplishing this step is infinite but, whatever the nature of the transformation may be, it always introduces some distortion.

In mathematical terms a map projection can be defined as a one-to-one correspondence between points on a datum surface (the generating globe) and points on a projection surface (a plane). A point on the generating globe is uniquely defined by two spherical coordinates: longitude λ and latitude ϕ. In the projection plane a point can be referenced by rectangular coordinates (x, y) or polar coordinates (r, Θ). With rectangular coordinates the mathematical relationship between both coordinate systems can be expressed by:

$$x = Rf_1(\lambda, \phi)$$
$$y = Rf_2(\lambda, \phi) \tag{1.2}$$

The functions f_1, f_2 define a unique projection system and determine the characteristics of the projection. It is usually assumed that f_1 and f_2 are real, single valued, continuous and differentiable functions of ϕ and λ in a certain domain. Hence the Jacobian determinant $J(x, y)$ may not vanish:

$$J(x, y) = \frac{\partial x}{\partial \phi} \frac{\partial y}{\partial \lambda} - \frac{\partial y}{\partial \phi} \frac{\partial x}{\partial \lambda} \neq 0 \tag{1.3}$$

1.2   SOME ELEMENTS OF GAUSS' THEORY

Let us consider a linear element of infinitesimal length $ds = pq$ on the generating globe. The intersections of the parallels ϕ and $\phi + d\phi$ and the meridians λ and $\lambda + d\lambda$ through p and q form an infinitesimal convex spherical quadrangle (figure 1.2a) which can be taken as a flat infinitesimal rectangle. Hence, we can assume ds to be an infinitesimal straight line segment that is defined as:

$$ds = \sqrt{(Rd\phi)^2 + (R\cos\phi d\lambda)^2} \tag{1.4}$$

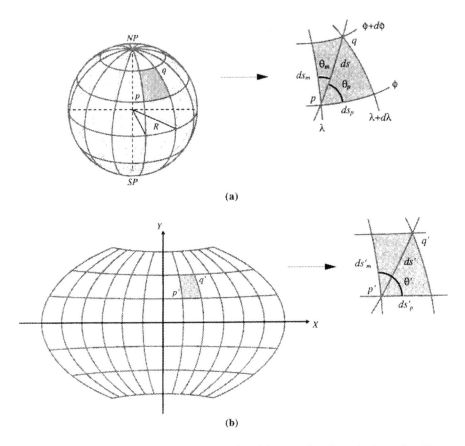

Figure 1.2 Elementary quadrangle on the generating globe (a), and on the projection surface (b).

The azimuth A of this line element, equal to θ_m, is the clockwise direction from the north and is given by:

$$\tan A = \frac{R\cos\phi d\lambda}{Rd\phi} = \cos\phi\frac{d\lambda}{d\phi} \tag{1.5}$$

If the quadrangle is mapped on a plane, it will be distorted. Generally, the direction and the length of the lines, as well as the angles between them, will be altered. Since the projected quadrangle (figure 1.2b) is also infinitely small, its sides and diagonals can be considered as straight lines. Applying Pythagoras' theorem:

$$ds'^2 = dx^2 + dy^2 \tag{1.6}$$

By differentiation of equation (1.2) and substitution in equation (1.6) one gets:

$$ds'^2 = Ed\phi^2 + 2Fd\phi d\lambda + Gd\lambda^2 \tag{1.7}$$

where E, F and G are the Gaussian fundamental quantities

$$E = \left(\frac{\partial x}{\partial \phi}\right)^2 + \left(\frac{\partial y}{\partial \phi}\right)^2$$

$$F = \frac{\partial x}{\partial \phi}\frac{\partial x}{\partial \lambda} + \frac{\partial y}{\partial \phi}\frac{\partial y}{\partial \lambda} \tag{1.8}$$

$$G = \left(\frac{\partial x}{\partial \lambda}\right)^2 + \left(\frac{\partial y}{\partial \lambda}\right)^2$$

1.2.1 Particular scales

The *particular scale* in any direction at any point μ is defined by the ratio of the projected length ds' over the original length ds:

$$\mu = \sqrt{\frac{Ed\phi^2 + 2Fd\phi d\lambda + Gd\lambda^2}{(Rd\phi)^2 + (R\cos\phi d\lambda)^2}} \tag{1.9}$$

Generally the particular scale varies from point to point and is different in every direction. This can be shown explicitly by substituting the expression for the azimuth (equation (1.5)) in equation (1.9):

$$\mu^2 = \frac{E}{R^2}\cos^2 A + \frac{G}{R^2\cos^2\phi}\sin^2 A + \frac{2F}{R^2\cos\phi}\sin A\cos A \tag{1.10}$$

The particular scales along the meridian and along the parallel, denoted h and k, are thus given by:

$$h = \frac{\sqrt{E}}{R} \tag{1.11}$$

$$k = \frac{\sqrt{G}}{R\cos\phi} \tag{1.12}$$

Since μ varies with direction it is interesting to investigate in which directions the particular scale attains its extrema. Using equations (1.11) and (1.12) and putting

$$p = \frac{2F}{R^2\cos\phi} \tag{1.13}$$

equation (1.10) simplifies to:

$$\mu^2 = h^2 \cos^2 A + k^2 \sin^2 A + p \sin A \cos A \tag{1.14}$$

With the condition for an extremum

$$\frac{d}{dA}\left(\mu^2\right) = 0$$

it is found that

$$\tan 2A_M = \frac{p}{h^2 - k^2} \tag{1.15}$$

indicating two orthogonal directions A_M and $A_M + 90°$ in which the maximum and minimum particular scales occur. These directions are called the *principal directions*.

1.2.2   Conformal projections

Since particular scales vary with direction, angles will be distorted as well, and meridians and parallels will not necessarily remain orthogonal after projection. It can be shown that the transformed angle θ' between meridians and parallels is given by (see Canters and Decleir, 1989, pp. 6–7):

$$\tan \theta' = \frac{\sqrt{EG - F^2}}{F} \tag{1.16}$$

or alternatively:

$$\sin \theta' = \sqrt{\frac{EG - F^2}{EG}} \tag{1.17}$$

It is clear from equation (1.17) that the orthogonality of meridians and parallels can be preserved by putting $F = 0$. Equation (1.15) indicates that for these so-called *orthogonal* projections the principal directions are situated along meridians and parallels.

When particular scales are independent from direction in every point of the map no angular distortion will occur and the projection is said to be conformal. Equation (1.10) shows that to accomplish this, in addition to $F = 0$, the following condition should be fulfilled:

$$\frac{E}{R^2} = \frac{G}{R^2 \cos^2 \phi} \tag{1.18}$$

1.2.3 Equal-area projections

The *area scale* σ for an arbitrary location on the map is given by the ratio of the area of the mapped quadrangle $ds'_m\, ds'_p \sin\theta'$ over the area of the quadrangle on the generating globe $ds_m\, ds_p$. Hence, it can also be written as:

$$\sigma = hk \sin\theta' \tag{1.19}$$

or with equations (1.11), (1.12) and (1.17):

$$\sigma = \frac{\sqrt{EG - F^2}}{R^2 \cos\phi} \tag{1.20}$$

A projection is said to be equal-area (or *equivalent*) if $\sigma = 1$ in every point on the map. The general condition for an equal-area projection therefore becomes:

$$\sqrt{EG - F^2} = R^2 \cos\phi \tag{1.21}$$

1.3 TISSOT'S INDICATRIX

Tissot (1881) demonstrated that at an infinitesimal scale every map projection is an affine transformation. If one calculates, in the point p on the sphere, the particular scale μ for each direction and then plots the computed values in the corresponding directions from the point p' on the map, the obtained locus of points is an ellipse, which is called the indicatrix of Tissot (figure 1.3) (for mathematical proof, see Canters and Decleir, 1989, p. 8). By choosing $ds = 1$ the indicatrix is also the representation in the plane of an infinitesimal circle with radius 1 on the generating globe. The semi-diameters a and b of the ellipse correspond with the maximum and minimum particular scales respectively, and hence with the principal directions (see section 1.2.1), which prove to remain orthogonal after projection.

From the definition of the particular scale (see section 1.2.1) and from figure 1.3 it is found that the particular scale in an arbitrary direction α is given by:

$$\mu = \sqrt{a^2 \cos^2\alpha + b^2 \sin^2\alpha} \tag{1.22}$$

If α_p is the direction of the parallel with respect to the principal X-axis, and $\alpha_m = \alpha_p + \pi/2$ is the direction of the meridian, then the particular scales along parallels and meridians, k and h, are given by:

$$k^2 = a^2 \cos^2\alpha_p + b^2 \sin^2\alpha_p \tag{1.23}$$

$$h^2 = a^2 \cos^2\alpha_m + b^2 \sin^2\alpha_m \tag{1.24}$$

Due to the orthogonality the latter equation can also be written as:

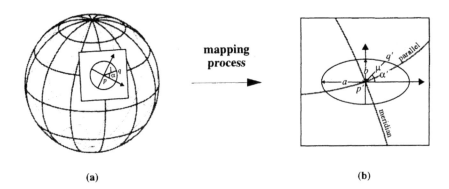

(a)   **(b)**

Figure 1.3 Elementary circle and principal axes in the tangent plane on the sphere (a), and corresponding ellipse of distortion in the mapping plane (b).

$$h^2 = a^2 \sin^2 \alpha_p + b^2 \cos^2 \alpha_p \tag{1.25}$$

Equations (1.23) and (1.25) combine to:

$$h^2 + k^2 = a^2 + b^2 \tag{1.26}$$

As will be shown later, equation (1.26) is very useful for determining the *principal scale factors a* and *b*.

Since the particular scale changes with direction, it is obvious that the resulting angular distortion is also direction dependent. It can be shown that a maximum change in direction Ω occurs for:

$$\tan \alpha_{max} = \pm \sqrt{\frac{a}{b}} \tag{1.27}$$

and is given by:

$$\sin \Omega = \frac{a - b}{a + b} \tag{1.28}$$

As illustrated in figure 1.4 the maximum change in direction Ω has an opposite sign in each two adjacent quadrants. This leads to a *maximum angular distortion* 2Ω for the angles subtended by the directions of maximum distortion in two adjacent quadrants, that is given by:

$$2\Omega = 2 \arcsin \frac{a - b}{a + b} \tag{1.29}$$

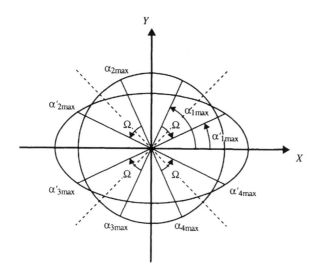

Figure 1.4 Superposition of elementary circle (generating globe) on the indicatrix (mapping plane) showing maximum change in direction Ω in the four quadrants, numbered 1–4.

From (1.29) it follows that the condition for conformality is now given by $a = b$, which implies that Tissot's indicatrix is a circle and that the particular scale is equal in all directions. The condition $a = b$ is identical with the previously given condition for conformality $h = k$ and $\theta' = \pi/2$ (see section 1.2.2).

The area scale is easily found by dividing the area of the indicatrix by the corresponding area of the infinitesimal circle:

$$\sigma = \frac{\pi ab}{\pi} = ab \tag{1.30}$$

If $ab = 1$ throughout the entire area of the map the projection will be equal-area. It is clear that conformality ($a = b$) and maintenance of correct area ($ab = 1$) are two mutually exclusive properties. Otherwise no distortion at all would occur, which is mathematically impossible since a sphere cannot be developed in a plane.

Equations (1.19) and (1.30) combine to:

$$ab = hk \sin \theta' \tag{1.31}$$

As will be shown immediately, this equation is very useful for the practical analysis of local distortion characteristics.

Analysing local distortion characteristics using Tissot's indicatrix

Tissot's indicatrix provides a full description of local map projection distortion. As shown above, all important distortion characteristics can be related to the principal scale factors a and b. Both a and b can be obtained by solving the following set of equations:

$$h^2 + k^2 = a^2 + b^2 \tag{1.26}$$

$$ab = hk \sin \theta' \tag{1.31}$$

where h, k and θ' are first calculated from the partial derivations of x and y with respect to ϕ and λ, equations (1.11), (1.12) and (1.17). Once a and b are known, maximum angular distortion and area scale can be calculated from equations (1.29) and (1.30) respectively.

To obtain a visual impression of the distortion characteristics of a map projection Tissot's indicatrix can be plotted at various locations on the map. To be able to do so, it is not only necessary to calculate the principal scale factors a and b. One also has to know the orientation of the indicatrix with respect to the map's coordinate system. For orthogonal projections, with the principal scale factors occurring along parallels and meridians, the positioning of the indicatrix is easily obtained. All one has to do is check in which direction the particular scale is the largest, along the parallels or along the meridians. This will indicate the position of the major axis of the indicatrix. Generally, however, parallels and meridians are not situated along the principal directions, and the angle between the X-axis of the projection and the major axis of the indicatrix θ'_a will have to be calculated.

To determine the value of θ'_a first the angle α'_p between the major axis and the parallel is calculated. From (1.23) it is found that:

$$\sin^2 \alpha_p = \frac{k^2 - a^2}{b^2 - a^2} \tag{1.32}$$

with α_p the angle between the direction of the major axis and the parallel on the generating globe. From the definition of the particular scale (see section 1.2.1) and figure 1.3 it follows that:

$$b \sin \alpha_p = k \sin \alpha'_p \tag{1.33}$$

hence:

$$\sin \alpha'_p = \sqrt{\frac{1 - a^2/k^2}{1 - a^2/b^2}} \tag{1.34}$$

The sign of α'_p will depend on the transformed angle θ' between the parallel and the meridian. If $\theta' > \pi/2$ (figure 1.5a) α'_p will be positive, else it will be negative (figure 1.5b). Once α'_p is known θ'_a can be calculated as follows:

$$\theta'_a = \theta'_p - \alpha'_p \tag{1.35}$$

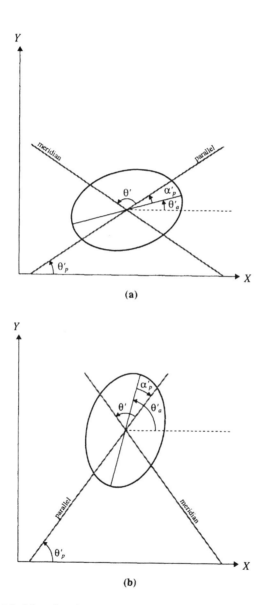

Figure 1.5 Orientation of the indicatrix for $\theta' > 90°$ (a), and for $\theta' < 90°$ (b).

with θ'_p the angle between the X-axis of the map's coordinate system and the parallel (figure 1.5). It is easy to show that θ'_p is given by:

$$\tan \theta'_{p} = \frac{\frac{\partial y}{\partial \lambda}}{\frac{\partial x}{\partial \lambda}} \tag{1.36}$$

Figure 1.6 shows Mercator's cylindrical conformal projection with indicatrices for various graticule intersections. Parallels and meridians are shown for 30° increments in latitude and longitude, starting from the equator and the central meridian. Since Mercator's projection is conformal all indicatrices are circles. However, as can be seen, the area of the circles increases with increasing latitude, indicating a strong variation in principal and areal scale factors. On the equator, which is a line of no distortion, the indicatrices are circles with unit radius. At a latitude of 60° the principal scale factor already reaches a value 2, which means that all distances at this latitude are doubled in length, while all areas are exaggerated by a factor 4.

Figure 1.7 shows Tissot's indicatrices for Behrmann's cylindrical equal-area projection with standard parallels at 30° latitude (Behrmann, 1909). Since the standard parallels are lines of no distortion, the indicatrices at 30° latitude are circles with unit radius. For other latitudes indicatrices are no longer circles, which means that particular scales vary with direction. Since cylindrical projections have an orthogonal graticule the axes of the indicatrices coincide with the parallels and meridians. In the lower latitudes, that is, in the area bounded by the two standard parallels, maximum scale exaggeration occurs along the meridians. In the higher latitudes maximum stretching occurs along the parallels. The stretching in both zones is compensated by an equal reduction of scale in the perpendicular direction. Hence all ellipses have the same area as the unit circle, and the projection is equal-area.

Figure 1.8 shows the graticule of the Winkel–Tripel projection (Winkel, 1921), an example of a map projection with a non-orthogonal graticule, and no special distortion properties (neither conformal, nor equal-area). This time the axes of the indicatrices no longer coincide with the directions of parallels and meridians, except at the equator and the central meridian, where parallels and meridians are perpendicular to one another. The eccentricity as well as the area of the indicatrices increases towards the edges of the graticule, indicating an increase in maximum angular distortion and areal scale factor. However, the eccentricity of the indicatrices near the edges of the projection is less pronounced than on Behrmann's equal-area graticule, while the variation in area scale is less than on Mercator's conformal projection. This illustrates the well-known fact that projections without any special distortion properties are often preferred to conformal or equal-area graticules for having a more balanced pattern of distortion. In the next chapter, dealing with the numerical evaluation of map projection distortion, this fundamental idea will be further explored.

Instead of showing Tissot's indicatrices for various locations on the map, different local measures describing one particular aspect of distortion such as, for example, maximum angular distortion, areal scale ratio, or principal scale factors, can be mapped by means of *isocols*, i.e. lines of equal distortion. Although isocols were already presented for simple projections in a number of German books and papers that were published at the end of the nineteenth century (see, for example,

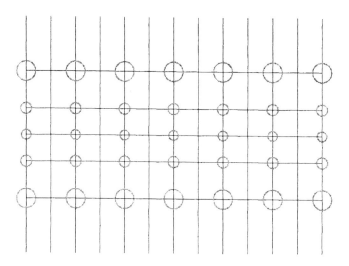

Figure 1.6 Indicatrices on Mercator's cylindrical conformal projection.

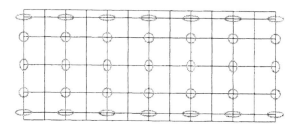

Figure 1.7 Indicatrices on Behrmann's cylindrical equal-area projection.

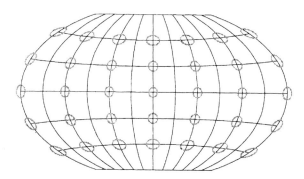

Figure 1.8 Indicatrices on the Winkel–Tripel projection.

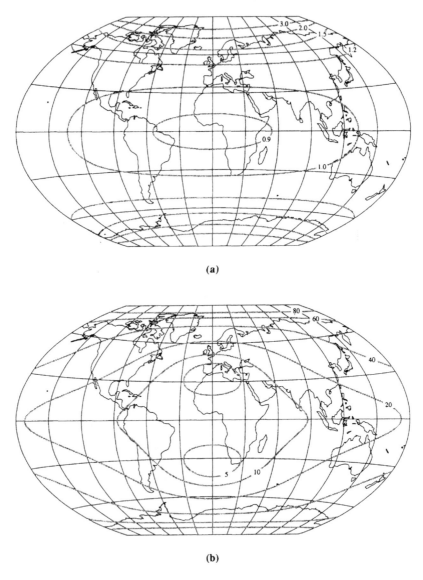

(a)

(b)

Figure 1.9 Lines of constant area scale (a), and lines of constant maximum angular distortion (b) for the Winkel–Tripel projection.

Hammer, 1889a; Zöppritz and Bludau, 1899, p.121, 148), it is probably Behrmann who was the first to present lines of constant distortion for various types of map projections. In a comparative study of fifteen equal-area projections, he calculated maximum angular distortion for small increments of latitude (10°) and longitude

(20°), constructed lines of constant maximum angular distortion for 0°, 1°, 5°, 10°, 20°, ..., and measured the area between successive lines in order to derive a mean maximum angular distortion value for the entire graticule (Behrmann, 1909) (see also section 2.2). Distortion patterns for all projections are included in the original paper. It should be mentioned that Behrmann's procedure of calculating a mean distortion value had already been suggested and applied in a slightly different way by Hammer (1889b) for choosing an optimal map projection for the African continent.

Figure 1.9 shows lines of constant area scale and constant maximum angular distortion, again for the Winkel–Tripel projection (see also figure 1.8). As can be seen, the area scale varies from less than 0.9 near the equator to more than 3.0 in the higher latitudes, meaning that most of the equatorial areas are slightly compressed, while polar areas are strongly exaggerated. In addition, angular distortion is moderate near the equator and increases towards the edges of the map, although the pattern of distortion is somewhat different from the previous one. From this single example it is clear that isocols are very useful for evaluating the distribution of distortion. In contrast with Tissot's indicatrices, isocols make it easier to obtain a quantitative impression of map projection distortion. This is probably one of the reasons why the drawing of isocols is usually preferred. It must, however, be said that most reference works on map projections only occasionally show patterns of distortion, in spite of their importance for a proper understanding and use of map projections. Also, in more recent studies the use of lines representing equal values of distortion is limited, although present computers make it relatively easy to construct them. Apart from Canters and Decleir (1989), until today no reference work is known in which all map projections are shown with their patterns of distortion.

1.4   CLASSIFICATION OF MAP PROJECTIONS

From the general definition of a map projection (see section 1.1) it is clear that the number of possible representations of the Earth on a plane is unlimited. For world maps alone hundreds of projections have been devised, although many of them have little practical significance. Also many so-called new projections are merely slight modifications of already existing ones while, due to the bad habit of designating a projection by the name of its author, the relation of the modification to the parent projection often becomes indistinct. Therefore, in order to gain more insight into the large variety of map projections and the relations that may exist among them, a rational classification seems necessary. Map projection classification may be of considerable practical value once a suitable projection has to be selected for a particular purpose, especially if the criteria for the grouping coincide with those used in the selection process. As will be shown, criteria for classification may relate to the geometry of the graticule or to specific properties of the projection.

A practical classification scheme should meet the following requirements:

1.   The number of classes should be restricted while at the same time the system must show enough flexibility to include an infinite number of possible projections.

2.   Overlapping of classes should be limited.

3.   The classification scheme should facilitate the selection of a projection for a specific purpose.

1.4.1   Geometric classification

Traditionally map projections are grouped into an *azimuthal*, a *cylindrical* and a *conical* class, according to the resemblance of their distortion patterns with those obtained by perspective projection of the sphere on a tangent plane, cylinder or cone (figure 1.10):

1. *Azimuthal* projections have one point of zero distortion (corresponding with the point where the plane touches the globe). Distortion increases radially from this central point leading to a distortion pattern consisting of concentric circular isocols. The characteristic graticule of an azimuthal projection is likewise circular. In the normal aspect (i.e. when the plane touches the globe at one of the poles) meridians are concurrent straight lines at angles equal to the difference between the corresponding longitudes, and parallels are concentric circles with the pole as their centre.

2. *Cylindrical* projections have a single, straight line of zero distortion (corresponding with the great circle where the cylinder touches the globe). Distortion increases in a direction perpendicular to and away from this so-called standard line. The characteristic distortion pattern thus consists of parallel, rectilinear isocols. In the normal aspect (i.e. when the cylinder touches the globe at the equator) parallels and meridians form an orthogonal rectilinear graticule, with the parallels equally divided by the meridians. The poles are then represented as straight lines, equal in length to the equator that has its correct size.

3. *Conical* projections have a single line of zero distortion (corresponding with the small circle where the cone touches the globe) which is shown on the map as a circular arc. Distortion increases in a direction perpendicular to and away from this standard circle. Hence the lines of constant distortion form a concentric pattern. The graticule of a conical projection is fan shaped. In the normal aspect meridians are represented as concurrent straight lines. The angles between the meridians on the projection and on the globe are proportional. Parallels are represented as concentric circular arcs. The parallel that coincides with the line of zero distortion is called the standard parallel. The centre of the parallels is not necessarily the pole. In general the pole is also shown as a circular arc, just like the other parallels.

While some map projections of the azimuthal, the cylindrical and the conical type are indeed obtained by perspective projection (and are therefore called *perspective* projections), for most map projections belonging to one of these three classes the spacing of the parallels is derived from differential calculus. This is in order to obtain favourable distortion properties such as, for example, no distortion of angles (see section 1.2.2), no distortion of area (see section 1.2.3), or minimum

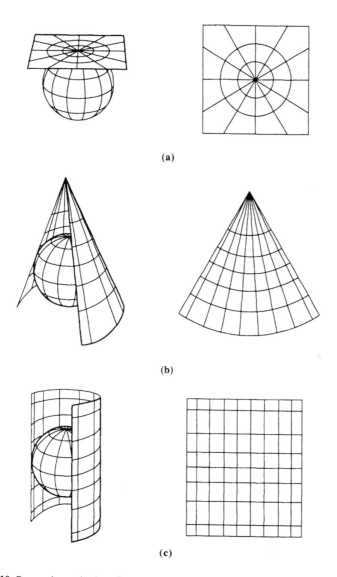

(a)

(b)

(c)

Figure 1.10 Perspective projection of the sphere on a tangent plane (a), on a tangent cone (b), and on a tangent cylinder (c).

overall distortion (see section 5.1). For these map projections the term azimuthal, cylindrical or conical only refers to an arrangement of the graticule that resembles the one obtained by projecting the Earth on a tangent plane, cylinder or cone respectively. The system for the spacing of the parallels on these map projections cannot be reproduced by perspective projection.

In the definitions of the azimuthal, cylindrical and conical projections, there is mention of the normal aspect of the three types of projections, where the standard point or standard line respectively correspond with one of the poles, with the equator or with a selected parallel, as indicated on figure 1.10. It is obvious that one can also produce azimuthal, cylindrical and conical maps by choosing the position of the point or line of no distortion in a more arbitrary way. Referring to the geometrical analogy we have been using so far, this can be accomplished by a re-orientation of the mapping plane or mapping body with respect to the Earth's surface. By doing so, the characteristic pattern of distortion of the projection is not changed. It is only centred on another part of the globe. Such a repositioning is referred to as a *change of aspect* (the term appears to have been introduced by Lee in 1944), and is mostly applied to place an arbitrary region of interest in the least distorted area of the map. While it has no impact on the nature of the distortion pattern itself, the appearance of the graticule, and therefore the representation of the continents, may be strongly altered. Depending on the position of the standard point or standard line, special names are given to particular aspects. In the next chapter all possible aspects of a projection will be treated in detail (see section 2.6). At this point it suffices to say that different aspects of a projection are not considered as separate projections. When map projections are classified according to the geometry of their graticule, reference is made only to the normal aspect of the projection.

For the sake of completeness it should be added that it is also possible to project the Earth on a cylinder, cone or plane that intersects the globe. The distortion pattern will then show two lines of zero distortion (instead of one) for the cylindrical and conical projections, and one standard circle (instead of a point) for the azimuthal projections. By adjusting the spacing of the parallels also non-perspective projections of the secant type can be defined. Again, however, for classification purposes these *secant* projections are usually not considered as being different from their tangent counterparts. Mathematically speaking they can be obtained by modifying the formulas of tangent projections in an attempt to reduce overall distortion (see section 4.1.1).

The basic geometric classification of map projections into an azimuthal, cylindrical and conical class has the advantage of simplicity, yet it is not all-inclusive. Literally hundreds of projections have been invented with graticules that do not fit the simple perspective model. Still most classification schemes that have been proposed in the last century are based on geometric classification, with new classes added to account for the large variety of existing graticules. An important classification is the one proposed by Lee (1944). Since azimuthal and cylindrical projections both can be considered as limiting cases of the conical projection, with the cone touching the sphere at the pole or at the equator, Lee places the three classes in one conical group. To accommodate other graticules, he adds three other classes – the *pseudocylindrical*, the *pseudoconical* and the *polyconical* class – of which the first and the third were defined already by Tissot (1881), and that can be considered as modifications of the cylindrical and conical projections (table 1.1):

Pseudocylindrical projections have straight parallels (as in cylindrical projections). The length of the parallels is variable so that the meridians are generally represented as curved lines that do not intersect the parallels at straight angles.

Pseudoconical projections represent the parallels as concentric circular arcs (as in conical projections). The length of the parallels may be individually adjusted so that the meridians are no longer shown as straight curves.

Polyconical projections have non-concentric circular arcs for parallels, as if the globe is projected onto a series of cones touching or cutting the globe at different latitudes. The centres of the parallels are located along the straight central meridian. Other meridians are represented as curves.

Some ambiguity exists with respect to the exact definition of the three classes. The general definitions above correspond to those given by Lee (1944). Yet class definitions are sometimes restricted to map projections on which the parallels are equally divided by the meridians, as it is the case for cylindrical and conical projections, and for which the graticule is symmetrical about a straight central meridian. Snyder (1993, p. 189), for example, speaks of *true* pseudocylindrical projections only if the meridians are equally spaced along each parallel. As such he does not consider the oval projections of the sixteenth century (see section 2.5.2) as *true* pseudocylindrical projections. For the pseudocylindrical and the polyconical class some authors also assume symmetry about the equator, since this is so for most well known projections belonging to these classes. Most authors, however, are less strict in their definitions. Projections with straight parallels and curved meridians, that do not satisfy the definition of a pseudocylindrical projection in the narrow sense, are usually considered as part of the same class. While many authors in their definitions still refer explicitly or implicitly to symmetry about a straight central meridian, the equal division of the parallels or the symmetry about the equator is generally not mentioned nor assumed.

All projections that do not belong to one of the six conical classes are placed in a second group of non-conical projections. Here Lee distinguishes *retroazimuthal*, *orthoapsidal* and *miscellaneous* projections. This time the distinction is not based on graticule geometry. The first class refers to projections with the azimuth from every point on the map to the centre point or to any point along the central meridian shown correctly (Littrow, 1833; Craig, 1910; Hammer, 1910). The second class consists of double projections that are obtained by drawing a parallel and meridian system on any suitable solid other than the globe and then making an orthographic projection of that (Raisz, 1943). Both classes include only few projections. The third class, which is called *miscellaneous*, includes all projections that cannot be conveniently accommodated elsewhere within the system.

Maling (1968) criticises Lee's classification scheme as well as other attempts to classify map projections for including classes like *miscellaneous*, *conventional* or *others*, arguing that the use of such a category creates "... a kind of rubbish bin ..." and "... represents a negative or despairing approach to the problems of classification". Similarly he pleads against the use of classes that are prefixed by "not-" or "non-", like the *non-conical* group and the *non-perspective* azimuthal class in Lee's classification. Of course, the reason for Lee's inability to devise an all-inclusive classification system, solely based on graticule geometry, lies in the fact that he starts with the three basic classes (azimuthal, cylindrical, conical

Small-scale map projection design

Table 1.1 Lee's geometric classification scheme (Lee, 1944).

CONICAL PROJECTIONS		
Conformal	Authalic	Aphylactic
CYLINDRIC		
Mercator	Authalic cylindric	Equidistant cylindric
		Rectangular cylindric
		Central cylindric
		Gall
PSEUDOCYLINDRIC		
	Collignon	Trapezoidal
	Sinusoidal	Apianus
	Mollweide	Loritz
	Authalic parabolic	Orthographic
	Prépetit Foucaut	Fournier II
	Eumorphic	Arago
CONIC		
Conformal conic with one standard parallel	Authalic conic with one standard parallel	Equidistant conic with one standard parallel
Conformal conic with two standard parallels (Lambert)	Authalic conic with two standard parallels (Albers)	Equidistant conic with two standard parallels
		Murdoch I, II, III
PSEUDOCONIC		
	Bonne	
	Sinusoidal	
	Werner	
POLYCONIC		
Lagrange	Authalic polyconic	Equidistant polyconic
Stereographic		Rectangular polyconic
		Fournier I
		Nicolosi
		van der Grinten
AZIMUTHAL		
Perspective		
Stereographic (sphere)		Orthographic
		Gnomonic
		Clarke
		James
		La Hire
		Parent
		Lowry
Non-perspective		
Stereographic (spheroid)	Authalic azimuthal	Equidistant azimuthal
		Airy
		Breusing

NON-CONICAL PROJECTIONS		
Conformal	Authalic	Aphylactic
RETROAZIMUTHAL		Equidistant retroazimuthal
ORTHOAPSIDAL	Authalic orthoapsidal pp.	Orthoapsidal pp.
MISCELLANEOUS	Aitoff	Schmidt
Littrow		Petermann
August		Two-point azimuthal
Peirce		(orthodromic)
Guyou		Two-point equidistant
Adams pp.		
Laborde pp.		

projections), which are strictly defined in terms of geometry. He then adds other, more loosely defined classes to accommodate existing projections that do not fit in the scheme. The only way to succeed in developing an all-inclusive scheme is to reverse the process and start with an initial scheme that contains only a few very general classes representing all possible configurations of the graticule. This scheme may then be further refined to an appropriate level to accommodate more detailed graticule definitions that may include the well-known geometrical classes described above.

1.4.2 Classification by the shape of the normal graticule

Inspired by Linnaeus' system of classification for plants and animals, Maurer (1935) developed a hierarchical classification scheme for map projections, consisting of *Stämme, Äste, Zweige, Ordnungen, Klassen, Unterklassen, Familien, Gattungen,* that is mainly based on the appearance of the meridians and parallels on the map. On a lower level also the principle of construction, the special properties of the projection, and the form of the map projection equations (algebraic, transcendental, ...) are taken into account. At the top of the hierarchy Maurer distinguishes five large categories (*Stämme*). The first category (*Stamm I*) includes all projections that have concentric circles or circular arcs for parallels. The second category (*Stamm II*) consists of all doubly symmetric graticules with straight parallels. Category three (*Stamm III*) groups all doubly symmetric graticules with non-straight parallels. In category four (*Stamm IV*) less regular projections are accommodated, i.e. projections that have no concentric circles or circular arcs for parallels, and/or just one axis of symmetry, or no symmetry at all. The fifth category (*Stamm V*) is added for non-continuous map projections, and includes interrupted graticules as well as graticules that are obtained by fusion of different parts of two or more continuous projections, using interruption or not. The technique has been introduced by Goode (1925), and since then has been applied quite a number of times, for non-interrupted (Erdi-Krausz, 1968; McBryde, 1978) as well as for interrupted map designs (Depuydt, 1983; Baker, 1986) (see also section 4.3).

Maurer developed his classification by partitioning a set of 237 map projections, almost all projections known at that time. The five main categories listed above are subdivided to a level of detail that responds to the variety of map projections to be distinguished. All six geometric classes from Lee's classification are also found in Maurer's scheme, although the terminology that is used to describe them is different. Each of the classes, however, is partitioned into a large number of subsets with more detailed definitions. To give an impression of the level of detail in Maurer's classification, table 1.2 shows the classification scheme for the second category (*Stamm II*), which includes the cylindrical and pseudocylindrical projections. All projection characteristics have been translated into commonly used English map projection terminology. The original terms used by Maurer to define the different hierarchical levels in his scheme have been kept.

Maurer's scheme is all-inclusive in the sense that (a) it contains all map projections the author was aware of, (b) it lends itself well to a further refinement if one should want to add new graticules. It represents the most comprehensive attempt to classification ever made. Unfortunately enough, the scheme is difficult to comprehend, not in the least because of the use of unfamiliar, germanised terms, of which the exact meaning is often hard to grasp, even for those who are familiar with the subject of map projection (see table 1.2). The scheme is too much detailed

Table 1.2 Part of Maurer's hierarchical classification scheme (*Stamm II*) (Maurer, 1935).

Ast A: equally spaced parallels

  Zweig A: exchangeable meridians (*fächerig*)

    Klasse I: cylindrical projections

      Unterklasse A: projections with pole line

        Familie A: perspective projections
        Familie B: equal-area projections
        Familie C: equidistant projections

      Unterklasse B: projections with pole at infinity

        Familie A: perspective projections
        Familie B: non-perspective conformal projections
        Familie C: non-perspective non-conformal projections

  Zweig B: pseudocylindrical projections (*nicht fächerig*)

    Ordnung a: projections with pole at infinity

      Klasse I: equal-area projections
      Klasse II: projections with equidistant parallels

Table 1.2 continued.

Ordnung b: projections which show the pole as a point or as a line

  Klasse 1: equally spaced parallels

    Unterklasse A: pointed-polar projections

      Familie A: all parallels have true proportion

        Gattung a: equidistant parallels
        Gattung b: non-equidistant parallels

      Familie B: equator and central meridian have correct proportion (2:1)

        Gattung a: straight meridians
        Gattung b: elliptical meridians

    Unterklasse B: projections with pole-line

      Familie A: arithmetic average of Plate Carrée and projections of *Unterklasse A*

        Gattung a: sinusoidal meridians
        Gattung b: straight meridians
        Gattung c: elliptical meridians

      Familie B: arithmetic average of cylindrical equidistant projection and projections of *Unterklasse A*

      Familie C: four parallels have true proportion

      Familie D: generalisations of *Familie A* and *Familie B* (Wagner, 1932)

Table 1.2 continued.

Klasse II: non-equally spaced parallels

    Unterklasse A: equal-area projections

        Familie A: projections with *a priori* defined outer meridian

            Gattung a: equal-area versions of *Klasse I, Unterklasse B, Familie A*

            Gattung b: equal-area versions of *Klasse I, Unterklasse A, Familie B*

            Gattung c: proj. with outer meridian obtained by averaging x-coordinate of cylindrical equal-area projection and sinusoidal projection

            Gattung d: equal-area versions of *Klasse I, Unterklasse B, Familie B*

            Gattung e: equal-area versions of *Klasse I, Unterklasse B, Familie D*

            Gattung f: projections with parabolic outer meridians

        Familie B: projections with *a priori* defined parallel spacing

            Gattung a: stereographic spacing of the parallels

        Familie C: affine transformations of *Familie A*

    Unterklasse B: non-equal-area projections

        Familie A: projections with elliptical outline and correct ratio of the axes

            Gattung a: orthographic spacing of the parallels

Ast B: non equally spaced parallels

    Zweig A: projections which show the pole as a point or as a line

        Klasse I: all meridians are circular arcs

    Zweig B: projections with pole at infinity

and too much based on purely theoretical considerations to be useful as a practical reference for map projection selection. In spite of Maurer's enormous effort his work is therefore of purely academic value. While it is often quoted in theoretical papers on map projection classification, most authors of reference books on map projection do not mention it, and certainly do not make use of it.

More recently Starostin *et al.* (1981) presented another all-inclusive classification scheme that is based on the shape of the normal graticule of meridians and parallels (for a treatise in English, see Bugayevskiy and Snyder, 1995, pp. 43–6). Although much less detailed and easier to comprehend than Maurer's scheme, it is very similar in concept, being also primarily based on the nature of the parallels and the symmetry level of the graticule. Starostin *et al.* distinguish two large sets of projections. The first includes projections with parallels of constant curvature, the second one groups all projections with parallels of various curvatures (table 1.3). The first set consists of three families, depending on whether the parallels are straight, concentric or eccentric circles (or circular arcs). Each family is subdivided into a number of classes, including the six geometric classes already found in Lee's classification, as well as the so-called *pseudoazimuthal* class, already defined by Tissot (1881). The latter consists of projections with concentric, circular parallels, just like the azimuthal projections, yet with curved meridians converging in the centre of the parallels. Only a very small number of projections of this type have been proposed. A distinction can be made between projections on which the meridians are all identical and equally spaced (a property that Maurer (1935) describes as *fächerig*), with the polar coordinate θ given by

$$\theta = \lambda + f_2(\phi) \tag{1.37}$$

and projections on which the meridians have different curvature and the polar coordinate θ is no longer a linear function of the longitude. The first pseudoazimuthal projection, which was developed by Wiechel in 1879, belongs to the first type and is of particular interest in the polar aspect since it combines the equal-area property with true scale along the meridians (Arden-Close, 1952). The projections developed by Bomford (Lewis and Campbell, 1951) and Ginzburg (1952) belong to the second, more general type (see also Snyder, 1993, pp. 242–4; Bugayevskiy and Snyder, 1995, pp. 129–32). Lee did not include the pseudoazimuthal class in his scheme, probably because there are only very few projections of this type known to exist. In Maurer's scheme (1935) map projections belonging to the pseudoazimuthal class are found in different parts of the classification, depending on the level of symmetry of their graticule.

As part of their first set of projections Starostin *et al.* (1981) also define a *cylindrical-conic* class, consisting of projections with straight parallels and concentric circular meridians, and two generalised classes, one of the cylindrical and one of the conic type, with the same properties as the original classes, but with variable spacing of the meridians (what Maurer would describe as *nicht fächerig*). For polyconic projections they make the distinction between polyconic projections *sensu stricto* and polyconic projections *in a broad sense*. The first class includes projections in which the radii of the parallels are derived by regarding each one as

Table 1.3 Starostin's classification scheme (Starostin *et al.*, 1981).

Subset 1: parallels of constant curvature

    Family 1: parallels are straight

        Class 1: cylindrical projections
        Class 2: generalised cylindrical projections
        Class 3: pseudocylindrical projections
        Class 4: cylindrical-conic projections

    Family 2: parallels are concentric circles

        Class 1: conic projections
        Class 2: generalised conic projections
        Class 3: pseudoconic projections
        Class 4: azimuthal projections
        Class 5: pseudoazimuthal projections

    Family 3: parallels are eccentric circles

        Class 1: polyconic projections "in a broad sense"
        Class 2: polyconic projections "in a narrow sense"

Subset 2: parallels of various curvatures

    Family 1: polyazimuthal projections

        Class 1: polyazimuthal projections "*sensu stricto*"
        Class 2: generalized polyazimuthal projections

    Family 2: generalized polyconic projections

        Class 1: parallels are elliptical
        Class 2: parallels are parabolic
        Class 3: parallels are hyperbolic
        Class 4: parallels of any curvature

    Family 3: polycylindrical projections

        Class 1: coordinates are given in an analytical form
        Class 2: coordinates are given in the form of a table

---

the standard parallel of a simple conic projection, the second class consists of projections in which the radii are not so derived. The term *polyconic* was applied to the first class by Hunt and Schott (1854), and extended to cover the second class by Tissot (1881) (see Lee, 1944). It should be added that Starostin *et al.* (1981) put no restrictions on the symmetry of the graticule in their definition of projection classes. To distinguish between graticules with different levels of symmetry, classes are split up into sub-classes, depending on the presence or absence of symmetry about equator, central meridian or both. This general definition of classes ensures the all-inclusiveness of the classification, yet it should be realised that it differs from the more restrictive definition of projection classes that is usually applied in map projection literature. The polyconic class, for example, is mostly restricted to graticules that are symmetrical about a straight central meridian. While almost all known projections of the polyconic type satisfy this constraint, there are some notable exceptions (see e.g. Bugayevskiy and Snyder, 1995, p. 148) that do not fit into the traditional definition of the class. Several map projections with less regular graticules, belonging to other classes, also have been developed, often with a special purpose (e.g. Poole, 1935; Siemon, 1935; Tobler, 1966a). All these projections can be included in Starostin's scheme.

In their second set of projections Starostin *et al.* also define three families (table 1.3). The family of so-called *polyazimuthal* projections represents the meridians by a set of straight or curved lines radiating from the pole, which is shown without interruption. Parallels are represented as ellipses (*polyazimuthal projections sensu stricto*) or lines of arbitrary curvature (*generalised polyazimuthal* projections), which distinguishes this family from the pseudoazimuthal projections on which parallels are concentric circles. The second and third family include graticules with an interruption near the pole, and with meridians and parallels represented by arbitrary curved lines. The distinction between both families is made in accordance with the form of the equations. For *generalised polyconic* projections equations are expressed in terms of polar coordinates. Four classes are defined, depending on the shape of the parallels (elliptical, parabolic, hyperbolic, and of any curvature). For the *polycylindrical* family of projections map position is expressed only in terms of rectangular coordinates. Two classes are distinguished: one with rectangular coordinates given in an analytical form, and one with coordinates given in the form of a table. Projections of the latter type have been proposed by Baranyi (1968) and Robinson (1974), and will be discussed in more detail in the third chapter of this study (see section 3.1). Just like in Starostin's first set of projections, the generalised polyconic and polycylindrical classes are divided into sub-classes depending on the presence or absence of symmetry about the equator and the central meridian.

1.4.3 Classification based on special properties

Although most map projection classification schemes are based on the geometry of the graticule, the preservation of special properties is often included as an additional criterion. This is not because map projection properties make it easier to distinguish between two map projections. On the contrary, while the identification of map projection class from graticule geometry is relatively easy, except for large-scale maps, most special properties are not deducible from the map's general appearance. Special properties, however, are often important in map projection selection. Some property may be especially useful or even indispensable for a particular application, yet the distortion patterns of projections having this property may be too unfavourable to make these graticules suitable for other use. This practical argument partly explains why special properties are often included in classification schemes.

For example, each main category in Lee's classification (table 1.1) is subdivided into a conformal, an *authalic* (equal-area) and an *aphylactic* class (neither conformal, nor equal-area) (Lee, 1944). As has been shown before, conformality and the preservation of area are two mutually exclusive properties (see section 1.3). As far as map distortion is concerned, conformal and equal-area projections can be considered as two extremes within the continuum of map projections (see section 2.4.2). Their application areas are well separated. While conformal projections are mostly used for large-scale applications, equal-area projections are usually preferred for the small-scale mapping of statistical distributions. *Aphylactic* projections occupy an intermediate position and are especially important for world maps or maps of very large areas.

Another well-known classification method that emphasises map projection properties is the one by Goussinsky (1951). He classifies projections by considering them from five different points of view (*classes*): nature, coincidence and position of the "projection surface" (what he calls the *extrinsic* factors), and properties and mode of generation of the projection itself (the so-called *intrinsic* factors) (table 1.4). While the five classes are not exclusive of each other, the *varieties* within each class are mutually exclusive. In class IV Goussinsky distinguishes three basic properties of map projections that are mutually exclusive: equidistancy, equivalency and *orthomorphism*, the term that used to be preferred by most British writers to describe conformality (Lee, 1944). As stated before equidistancy can only be achieved in one direction. When speaking of equidistant projections one usually refers to projections with no scale distortion along the meridians. Goussinsky states that all other properties of projections are of secondary nature. Apart from the three basic properties he makes no attempt to define other varieties of property.

In his fifth class (*mode of generation*) Goussinsky distinguishes between *perspective, projective,* and *conventional* projections. Alternatively he also speaks of *geometric, semi-geometric,* and *non-geometric* projections. For *perspective* projections, in which the sphere is geometrically projected from a fixed point of view, the graticule characteristics are solely based upon the position of this point with respect to the projection surface. While some of these projections may have a special property – for example, the stereographic azimuthal projection is conformal, the gnomonic azimuthal projection represents all great circles as straight lines – the practical use of these projections is limited because, as Goussinsky puts it, "... they are not sufficiently pliable". For *projective* projections, the overall appearance of the graticule is identical to purely perspective projections, only the spacing of the parallels is adjusted to satisfy a certain *a priori* imposed condition. All non-perspective azimuthal, cylindrical and conical projections that have one or more well-defined properties belong to this variety. Both perspective and projective projections are united in one category, which Goussinsky calls *regular* projections, as opposed to *conventional* or *irregular* graticules, which do not result, fully or partly, from a projection operation, but for which one is free to choose the form of both the parallels and the meridians.

Table 1.4 Goussinsky's classification scheme (Goussinsky, 1951).

Classes	Varieties		
	THE PROJECTION SURFACE (Extrinsic considerations)		
I NATURE	Plane	Conical	Cylindrical
II COINCIDENCE	Tangent	Secant	Polysuperficial
III POSITION	Direct	Transverse	Oblique
	THE PROJECTION IN ITSELF (Intrinsic considerations)		
IV PROPERTIES	Equidistancy	Equivalency	Orthomorphism	Other properties
V GENERATION	Perspective	Projective	CONVENTIONAL

Since irregular projections are not related to any projection surface, Goussinsky proposes to refer to these graticules as *representations* instead of *projections*. Also Lee (1944) had already criticised the general use of the term projection – which has long outgrown its geometrical meaning – before, yet without suggesting a proper alternative. In spite of this criticism the term *projection* is still used in the broader sense today. As Goussinsky's classification is strongly based upon the characteristics of the projection surface, it is only applicable to regular projections. According to Goussinsky each regular projection can be uniquely described by a set of varieties, one from each class. Also any combination of varieties, if theoretically possible, defines a unique projection. Hence the number of possible regular projections is limited and can be determined *a priori* by purely analytical considerations. On the other hand, the number of irregular projections is unlimited and will definitely increase with time. As such these projections are the subject of empirical or *a posteriori* classification (Goussinsky, 1951).

Next to equidistance, conformality, and the preservation of area there are many other properties that are relevant to map projection use, including representing all circles on the sphere as circles on the map, representing the orthodrome or loxodrome as a straight line, and so on (see section 6.2). Maling (1968) proposed a list of not less then eleven special properties that may be of value to the map user. Yet the list of possible properties has clearly no ending, and the complexity of a classification that is strongly based on projection properties would continue to grow as more properties are added. Some classes corresponding to more common properties (e.g. equidistant meridians) would contain many projections, while other classes representing more restrictive properties would only contain a single projection (e.g. straight orthodromes). Also the use of special properties as a criterion for classification will always create a large heterogeneous class of projections that do not possess one of the listed properties, and that may have very different characteristics. Lee's *miscellaneous* class (table 1.1) provides a good example.

Goussinsky seems to suggest the definition of additional varieties in class IV of his scheme (table 1.4), which is absolutely necessary to guarantee the claimed uniqueness of his classification. Yet he does not attempt to define these "other properties", and so cleverly avoids all practical problems raised by a property-based scheme. One of these problems that has not been mentioned yet is that not all properties are mutually exclusive, which makes it extremely difficult to avoid the presence of overlapping varieties. A unique classification scheme, as suggested by Goussinsky, can only be accomplished by defining all feasible combinations of map projection properties, after a meticulous study of what is theoretically possible. This would make the scheme less clear and more difficult to use.

From all that has been said it should be clear that projection properties do not provide a good basis for classification. Since they are of importance to the cartographer and the map user, some mutually exclusive properties can be part of a hierarchical scheme (Maurer, for example, refers to important special properties at various levels in his classification). Also map projection properties should definitely be referred to in the name of the projection (see section 1.4.5). Yet they are not suited as a primary criterion for classification.

1.4.4   Parametric classification

Both Maurer (1935) and Lee (1944) started from existing map projections in setting up their classification schemes. Starostin *et al.* (1981) extended the definition of existing classes, and added new ones to include all possible projections. Tobler (1962) took a more general approach and proposed a parametric classification consisting of four *groups* A–D, corresponding to the following mathematical relationships between the coordinates in the plane (u,v) and the geographical coordinates (λ,ϕ):

A	B	C	D
$u = f_1(\lambda,\phi)$	$u = f_1(\lambda)$	$u = f_1(\lambda,\phi)$	$u = f_1(\lambda)$
$v = f_2(\lambda,\phi)$	$v = f_2(\lambda,\phi)$	$v = f_2(\phi)$	$v = f_2(\phi)$

Distinguishing between rectangular coordinates (x,y) and polar coordinates (r,θ), these four groups define eight functional relationships that are illustrated schematically in figure 1.11. Some ambiguity may arise as to whether a projection should be defined in terms of rectangular or polar coordinates. In principle, every possible configuration of the graticule can be mathematically expressed with respect to both coordinate systems. As a rule, Tobler suggests the use of rectangular coordinates when the map is centred on the intersection of the equator and the central meridian. Polar coordinates should be used when the map is centred on one of the poles. Of course, Tobler refers to the normal aspect of the projection, i.e. the aspect for which there is a direct relationship between the graticule and the distortion pattern, both having the same axes of symmetry (see section 2.6). Since different aspects of the same projection are not to be considered as different projections, only the normal aspect of the projection is used as a criterion for the grouping.

Tobler's parametric classification is easy to understand and includes all possible map projections representing the Earth in a continuous fashion. Nevertheless, Tobler's four-fold subdivision does not go far enough to be practically useful. Groups C and D comprise projections that are very different in appearance even though they satisfy the same functional relationships. On the other hand, group B contains only a few projections, which are used for specific geographic purposes and are defined in terms of polar coordinates (Tobler, 1963). In conventional cartography these projections are seldom used. To obtain a more detailed scheme, Maling (1992) suggests a combination of Tobler's parametric classification and the traditional geometric approach, the first having the advantage of being all-inclusive, the second being well adapted to the present diversity of well-known and frequently used projections.

Maling (1992) distinguishes seven geometric *classes* (figure 1.12) that can easily be related to Tobler's major subdivision (figure 1.11). All of these classes have been mentioned before (see sections 1.4.1, 1.4.2). Yet one of them, the polyconic class, is given a different definition. While in most textbooks the polyconic class has non-concentric, circular arcs for parallels, and a straight central meridian, Maling extends the definition of this class to incorporate all projections with curved meridians and parallels. As such his polyconic class corresponds with Starostin's polyconic family, plus all projections in the second set of his classification (see section 1.4.2), and equals group A of Tobler's classification.

Within each class Maling makes a further distinction between projections with a constant, decreasing, or increasing separation of the parallels, proceeding from the equator towards the poles (table 1.5). Although the parallel spacing provides an additional clue for the identification of map projections, one might argue if it would not have been more interesting to consider the spacing of the parallels away from the standard line (or standard point) instead of away from the equator. This would have made the classification more useful for selection purposes. Indeed, the distortion characteristics of a projection as well as the most important special properties (equal-area property, conformality, and equidistance), are strongly related to the spacing of the parallels. Nearly all equal-area projections show a decreasing parallel spacing away from the standard line (or standard point). Conformal projections are characterised by an increased spacing of the parallels. The relation between the spacing of the parallels and the distortion characteristics of a map projection will be discussed in more detail in the next chapter (see section 2.5.1).

By introducing the spacing of the parallels as an additional criterion, Maling obtains an ordered hierarchy of *groups*, *classes* and *series*. Apart from Tobler's four major groups (A–D), and similar to Maurer's classification (1935), Maling also defines a fifth group (E) comprising all *composite* map projections, which cannot be described by a unique set of transformation formulas. While Tobler's parametric

Groups (Tobler)   **Classes (Maling)**

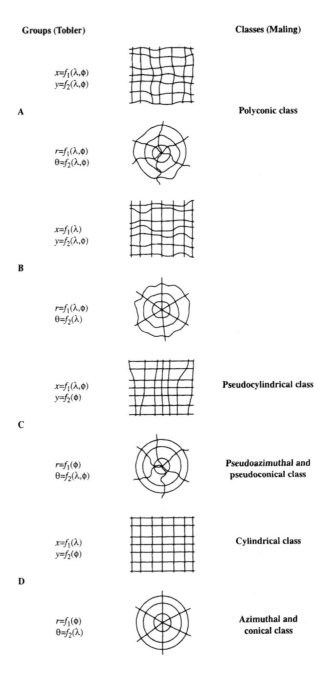

$x=f_1(\lambda,\phi)$
$y=f_2(\lambda,\phi)$

A   Polyconic class

$r=f_1(\lambda,\phi)$
$\theta=f_2(\lambda,\phi)$

$x=f_1(\lambda)$
$y=f_2(\lambda,\phi)$

B

$r=f_1(\lambda,\phi)$
$\theta=f_2(\lambda)$

$x=f_1(\lambda,\phi)$ Pseudocylindrical class
$y=f_2(\phi)$

C

$r=f_1(\phi)$ Pseudoazimuthal and
$\theta=f_2(\lambda,\phi)$ pseudoconical class

$x=f_1(\lambda)$ Cylindrical class
$y=f_2(\phi)$

D

$r=f_1(\phi)$ Azimuthal and
$\theta=f_2(\lambda)$ conical class

Figure 1.11 Tobler's parametric classification scheme with corresponding classes according to
Maling (after Tobler, 1962; Maling, 1992).

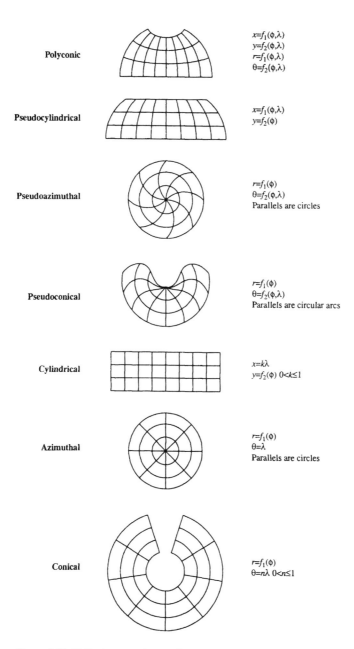

Figure 1.12 Maling's seven classes of map projections (after Maling, 1992)

Table 1.5 Maling's ordered hierarchy of groups, classes and series (Maling, 1992).

Group	Class	Series		
		Decreasing separation	Equidistant spacing	Increasing separation
D	CYLINDRICAL	Cylindrical equal-area (Lambert)	Plate Carrée	Mercator's projection Braun's perspective cylindrical
	AZIMUTHAL	Stereographic Gnomonic Minimum-error (Airy) projection Breusing's geometrical mean azimuthal Breusing's harmonic mean azimuthal	Azimuthal equidistant (Postel)	Azimuthal equal-area (Lambert) Orthographic
	CONICAL	Conical equal-area with truncated pole Conical equal-area with point pole (Lambert)	Equidistant conical (Ptolemy) Minimum-error conical (Murdoch III)	Conformal conical (Lambert)
C	PSEUDO-CYLINDRICAL	Mollweide's projection Pseudocylindrical equal-area with elliptical meridians (Fournier II) Parabolic	Sinusoidal Pseudocylindrical with elliptical meridians (Apianus II) Polyhedric	
	PSEUDOAZIMUTHAL PSEUDOCONICAL		Bonne's projection	Equal-area pseudoazimuthal (Wiechel)
A	POLYCONIC	Hammer–Aitoff projection	Simple polyconic Aitoff's projection Tripel projection (Winkel)	

classification is all-inclusive, one should realise that Maling's scheme is not. Maling's classification covers most frequently used map projections, and will therefore be suited for most practical purposes, yet it cannot accommodate all existing map projections, and certainly not all those that are theoretically possible. In the fifth chapter of this study, dealing with the development of low-error map projections, several graticules will be discussed that do not fit into Maling's seven classes, but can only be attributed to one of Tobler's major groups. Despite the fact that Maling himself criticises the use of classification criteria that are not all-inclusive (Maling, 1968), his subdivision in series ignores the presence of map projections with the spacing of the parallels increasing *and* decreasing as the latitude varies from equator to pole (see e.g. Baranyi, 1968).

1.4.5 Nomenclature of map projections

In the nomenclature, associated with the classification system, the name of each projection should reflect the appearance of the graticule. The naming should also be brief, easy to comprehend and unequivocal.

In the past projections were often named after their inventor. This has led to confusion since there is often disagreement on who is the original author of a projection. Moreover, several projections may be due to the same person. Very often it then concerns projections with minor differences in appearance. At best those projections are numbered, e.g. Eckert I to Eckert VI. Although such a numbering helps identification it appeals only to the memory. Numbering has clearly no relation to the appearance of the graticule or the special properties involved.

Maling (1968) proposed a series of descriptive terms that should lead to an unequivocal identification of a projection:

1. The projection class: this allows one to locate the graticule within the classification scheme and, by the same token recognition of the distortion patterns. If a hierarchical scheme is used, a full description of the various subclasses may be necessary to identify the exact position in the classification (e.g. referring to Maling's classification, *pseudocylindrical projection with equally spaced parallels*).

2. Special properties: they are relevant in relation to the potential use of the projection (e.g. conformal, equal-area, ...).

3. Distinctive characteristics: they provide additional information which should make a complete identification possible, such as the nature of modification, the characteristic outline of the projection, the general appearance of meridians, parallels or poles, etc. (e.g. *pseudocylindrical projection with equally spaced parallels, elliptical meridians and pole line*).

In some cases, however, even a lengthy description still fails to define a projection without ambiguity. Therefore it might be necessary to combine the descriptive terms with a reference to the author. For example, several pseudocylindrical projections with equally spaced parallels, elliptical meridians and pole line have been defined (see Eckert, 1906; Maling, 1960; Wagner, 1962; Snyder, 1977), so only reference to the author allows proper identification of the projection. Also for map projections with a long history (e.g. Mercator projection, Mollweide projection) the author should best be mentioned, this to avoid that the map reader who is accustomed to the use of certain names should become disorientated. The same applies for special references given by some authors to identify their own inventions (e.g. eumorphic projection (Boggs, 1929), flat-polar quartic (authalic) projection (McBryde and Thomas, 1949), and Atlantis projection (Bartholomew, 1948)).

Maling (1968) also rightly states that the aspect of the projection should be properly indicated. Reference to the aspect should be made for all transverse and oblique variations of a map projection, because the aspect affects the appearance of the graticule. In addition, it might be useful to include information on the projection centre, the orientation of the axes of the distortion pattern, ... For example, Bartholomew's Atlantis projection (figure 2.2) might be referred to as a transverse oblique Mollweide projection with origin at 45°N, 30°W (for a more detailed treatment of the different aspects of a map projection, see section 2.6).

CHAPTER TWO

Map projection distortion at the infinitesimal scale

In the first chapter of this study, the most important elements of map projection theory have been introduced. As has been demonstrated, according to Tissot's theory every map projection is locally an affine transformation. An infinitesimal circle surrounding a point on the globe is transformed into an infinitesimal ellipse, the so-called indicatrix of Tissot, surrounding the corresponding point on the image surface. Once the lengths of the axes of the indicatrix are known it is possible to determine all distortion characteristics of the projection in the immediate vicinity of the position for which they are calculated. Local distortion characteristics for a projection can be visualised by mapping Tissot's indicatrix for selected positions on the graticule or by constructing lines of constant distortion (see section 1.3). Although graphic representations of this kind offer a clear and detailed insight in the spatial distribution of map projection distortion, they do not allow a quick comparison of distortion for a large set of map projections. They also do not give us information about the optimal choice of parameter values (e.g. position of the centre, latitude of the standard parallels) for a particular projection. Both comparative evaluation of map projection distortion and optimisation of map projection parameter values require that distortion (or at least one aspect of it, e.g. angular, area or distance distortion) is quantified by a single measure.

This chapter deals with the definition and application of integral measures of distortion, derived from local distortion theory. The main objective of this part of the study is to investigate the usefulness of these measures for quantifying distortion on small-scale maps that cover large areas. While large-scale maps may be used for measuring angles, areas and distances, there is usually no need to do precise measurements on small-scale maps, and stress will lie on the reduction of the overall distortion of large shapes. Although local distortion theory is generally applied for evaluation purposes, in large-scale as well as in small-scale mapping, a review of earlier studies on the evaluation of small-scale map projection distortion indicates a certain inability to properly characterise the distortion of large shapes as it is perceived by the map user. Building on traditional evaluation techniques that have been suggested in cartographic literature, in this chapter integral measures of angular, area and distance distortion will be defined, and distortion will be analysed, locally and globally, for a large set of small-scale map projections. This will permit a detailed study of the impact of graticule geometry on the distortion pattern of a projection. It will also allow us to assess the utility and the limitations of a quantitative analysis of local distortion characteristics.

2.1   LOCAL MEASURES OF DISTORTION

As has been shown in the first chapter, the distortion of a map projection is the result of a variation of scale throughout the mapped area. The directional variation of scale leads to an angular distortion that is, of course, also dependent on direction. As shown on figure 1.4 it reaches its maximum value 2Ω for the angles submitted by the directions of maximum distortion. This value is mostly taken as a measure of angular distortion at the infinitesimal scale. It is given by:

$$2\Omega = 2\arcsin\frac{a-b}{a+b} \tag{2.1}$$

with a and b the scale factors along the principal directions.

It is clear that after projection the area of the indicatrix will generally be different from the area of the elementary circle. As explained before, the corresponding areal scale factor can easily be defined as the area of the indicatrix divided by the area of the elementary circle:

$$\sigma = ab \tag{2.2}$$

A measure of local areal scale error often used is:

$$e_{ar} = (ab-1)^2 \tag{2.3}$$

In 1861 the British geodesist George Biddell Airy proposed to measure the overall impact of scale distortion by means of a hybrid measure that quantifies the combined effect of angular and area distortion, both caused by the distortion of scale:

$$e_{AI} = \frac{1}{2}\left[\left(\frac{a}{b}-1\right)^2 + (ab-1)^2\right] \tag{2.4}$$

If scale distortion is independent from direction the first term becomes zero and the distortion of scale only causes a distortion of area. If the product of the scale factors in both principal directions is equal to one the second term vanishes and the scale distortion will lead to a distortion of angles only. In all other cases, both terms in equation (2.4) will contribute to local error. Airy proposed this measure of error to define a new azimuthal projection with a minimum "total evil", to be obtained by integration of e_{AI} over the entire map area and minimisation of the total error, applying the method of least squares (Airy, 1861). Since equation (2.4) includes an angular and an area component of error, Airy speaks of map projection optimisation "by balance of errors". Yet, he never applied his original criterion for optimisation purposes. Instead, he made use of the following measure of local error

$$e_{A2} = \frac{1}{2}\left[(a-1)^2 + (b-1)^2\right] \tag{2.5}$$

which is the mean of the squared scale errors along the principal directions. Airy wrongly assumed that $e_{A1} = 2e_{A2}$. Györffy (1990) demonstrates that both criteria are clearly different, with $e_{A1} - 2e_{A2}$ becoming larger as the scale error increases. This means that the use of (2.5) does not lead to a projection determined "by balance of errors", as Airy had originally intended it. Nevertheless, e_{A2} became one the most applied measures of local scale error, and a frequently used criterion for the development of minimum-error projections (see section 5.1). Györffy also shows that e_{A2} has a distinct geometrical meaning. It represents the mean squared deviation between a point on the elementary circle and the corresponding point on Tissot's indicatrix, obtained by integrating the squared deviation over the entire circle (2π). Hence, one might say that e_{A2} not only accounts for the scale error along the principal axes but, indirectly, for the scale distortion in all directions.

Many other measures have been proposed to characterise the distortion of scale in the immediate vicinity of a point. Jordan (1896) suggested integrating the squared scale error in an arbitrary direction $(\mu - 1)^2$ over the full azimuthal range:

$$e_J = \frac{1}{2\pi} \int_0^{2\pi} (\mu - 1)^2 \, d\alpha \tag{2.6}$$

Klingach generalised Airy's first distortion measure e_{A1} by introducing two weight factors P_ω and P_σ to establish any desired balance between angular and area distortion (see Baetslé, 1970; Bugayevskiy and Snyder, 1995, p. 26):

$$e_K = \left[P_\omega \left(\frac{a}{b} - 1 \right)^2 + P_\sigma (ab - 1)^2 \right] \Big/ (P_\omega + P_\sigma) \tag{2.7}$$

The squares in expressions (2.5) and (2.6) make that all terms in the summation are positive. This is a necessary condition to prevent enlargements and reductions from compensating each other. However, large errors will receive relatively more weight than small errors. In addition, enlargements and reductions by the same factor do not have equal weight in the summation. Both arguments indicate that Airy's and Jordan's measures are less suited for error evaluation and optimisation purposes if the scale factors strongly differ from unity, as it is the case with the small-scale mapping of large areas. Kavrayskiy (1958) modified Airy's and Jordan's measure by replacing the scale error $\mu - 1$ by the natural logarithm of the scale factor $\ln \mu$, leading to the so-called Airy–Kavrayskiy criterion:

$$e_{AK} = \frac{1}{2} \left[(\ln a)^2 + (\ln b)^2 \right] \tag{2.8}$$

and the Jordan–Kavrayskiy criterion:

$$e_{JK} = \frac{1}{2\pi} \int_0^{2\pi} (\ln \mu)^2 \, d\alpha \tag{2.9}$$

The use of the natural logarithm assures that enlargements and reductions by the same factor receive equal weight. Extreme distortion values now weigh relatively less than small distortion values, which reverses the effect obtained with squared measures. Still other, less popular measures of local distortion have been proposed by Weber (1867), Konusova (see Bugayevskiy and Snyder, 1995, p. 26), and others.

2.2   INTEGRAL MEASURES OF DISTORTION

In order to compare map projections with respect to their distortion characteristics, it seems appropriate to quantify overall distortion by one single measure. Tissot (1881) proposed to characterise the distortion of a map projection by the maximum value attained over the mapped area. Maximum distortion values can easily be calculated but they have one major disadvantage. They do not take into account how distortion is spatially distributed. Map projections with similar maximum distortion values may have very different patterns of distortion. Nevertheless, the method has often been used for map projection comparison, as well as for the optimisation of map projection parameters and the development of new projections. A well-known example is the work of Wagner (1962), who characterises the area distortion of a world map by the maximum value of the areal scale factor between the equator and the parallel of 60°, and states that this maximum value should never exceed 1.2. He subsequently developed several new pseudocylindrical map projections with the area distortion increasing from 0% at the equator to 20% at 60° latitude (see sections 2.5.2, 4.1.2). Peters (1975, 1978) adopted Wagner's measure to characterise the area distortion for a number of projections that are frequently used for world maps (see section 3.4).

To take into account the spatial variation of distortion, the local measures that have been defined above can be integrated over the region of interest. Dividing the obtained sum by the total area of the region S yields a mean distortion value that can be taken as a measure of overall distortion:

$$E = \frac{1}{S} \int_S e \, dS \qquad\qquad (2.10)$$

Using, for example, the local measures (2.1), (2.3) and (2.5), and eliminating the constant factor 0.5 in the latter, integral measures for the evaluation of the angular, area and scale distortion over a well-determined part S of the map are respectively given by:

$$E_{an} = \frac{1}{S} \int_S \left(2 \arcsin \frac{a-b}{a+b} \right) dS \qquad\qquad (2.11)$$

$$E_{ar} = \frac{1}{S} \int_S (ab - 1)^2 \, dS \qquad\qquad (2.12)$$

$$E_{ab} = \frac{1}{S} \int_S \left[(a-1)^2 + (b-1)^2 \right] dS \qquad (2.13)$$

The first two measures are zero for conformal and equal-area projections respectively. Since a sphere cannot be developed into a plane without distortion the value of the third criterion will always differ from zero. As explained before, the squares in (2.12) and (2.13) assure that all terms in the summation are positive.

In practice the mean distortion value can be calculated by dividing the region of interest into k small plots, determining the value of e for the midpoint of each plot, and averaging the obtained values, using the area of each plot S_k as a weight factor:

$$E = \frac{1}{S} \sum_{i=1}^{k} e_k S_k \qquad (2.14)$$

where:

$$S = \sum_{i=1}^{k} S_k \qquad (2.15)$$

Probably the first to suggest the calculation of mean distortion values to evaluate map projection distortion was Ernst Hammer (1889b). At the end of the nineteenth century, he became involved in a debate with Alois Bludau, another leading cartographer of the time, about the choice of an optimal equal-area projection for the African continent. While Bludau (1891, 1892) used the maximum angular distortion within the area to be mapped as a decisive criterion, Hammer (1889b, 1894) proposed a graphical method to determine the mean angular distortion ("durchschnittliche Maximalwinkelverzerrung") for equal-area projections. The method was based on the measurement of the area between successive lines of constant angular distortion drawn on the graticule of the projection, using a planimeter. The lines were constructed by calculating the maximum angular distortion value (equation (2.1)) for equal increments of longitude and latitude. For each zone between two isocols, the average of both delimiting isogram values was taken as the mean distortion value for that area. To obtain a mean angular distortion value for the entire graticule, the mean values for each zone were multiplied by the area of the zone, added up, and divided by the total area of all zones. Hammer applied the method to compare the overall angular distortion for a number of map projections, and decided that the oblique conical equal-area projection was best suited for a map of the African continent.

As pointed out by Behrmann (1909), Hammer's approach can only yield reliable results for a high density of isocols. If the interval between two subsequent lines of constant distortion is too large, the average of both delimiting distortion values will not describe the continuously varying angular distortion within that zone in a satisfactory way. To obtain more accurate results Behrmann slightly modified Hammer's method. After measuring the area between successive lines of constant distortion he constructed a hypsographic-like curve by assigning each zone, from the zone with the lowest to the zone with the highest distortion, to a part of the x-axis proportional to the area of that zone. He then represented the maximum angular distortion value for each line along the y-axis, and connected the obtained points by a smooth curve. By measuring the total area beneath the curve (again using a planimeter), and dividing this area by the base of the diagram, the mean angular distortion value is obtained. Behrmann applied his method to calculate the mean angular distortion for fifteen equal-area projections, for one hemisphere as well as for the entire globe. Figure 2.1 shows Behrmann's "hypsographic" distortion curve for Lambert's azimuthal equal-area projection, with mean distortion values for the hemisphere and the globe indicated on the y-axis. Table 2.1 lists the mean angular distortion values for all fifteen projections for the mapping of the entire globe. As can be seen, angular distortion is the lowest for the equal-area cylindrical projection with standard parallels at $\pm30°$ latitude. Behrmann considered this projection, which was named after him, as the best equal-area choice for a global representation of the Earth. For a suggestion of the curvature of the Earth's surface he recommends Eckert's equal-area pseudocylindrical projection with elliptical meridians and pole line (Eckert IV) (see also section 2.5.2), with a mean angular distortion value almost as low as for his cylindrical projection.

Hammer and Behrmann's methods both yield approximate values of (2.11). Their approach is rather complicated, is susceptible to errors that are introduced by the interpolation procedure and the measurement of area, and is only applicable to equal-area projections. Nowadays, with the availability of fast computing resources, the calculation of mean distortion values has become much easier, and the

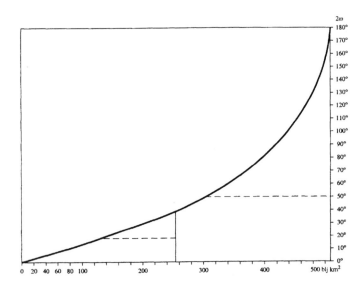

Figure 2.1 Behrmann's "hypsographic" distortion curve for Lambert's azimuthal equal-area projection (after Behrmann, 1909).

calculation of distortion values for a large number of points has made it possible to reach a high level of accuracy. It should however be noted that the evaluation of distortion, using equation (2.14), implies the exclusion of singular points, i.e. points where the one-to-one correspondence between the Earth and the map is not satisfied (e.g. poles that are represented by lines or that cannot be shown at all). For the evaluation of distortion on world map projections, or for the optimisation of map projection parameter values, the calculation of mean distortion values is often limited to latitudes between –85° and 85° (Francula, 1971; Györffy, 1990) or even smaller latitudinal ranges, to avoid that extreme distortion values weigh too heavily on the determination of the mean distortion value. This should not necessarily be interpreted as an important drawback of the evaluation procedure. Already in 1951, Robinson stated that the determination of a mean distortion value over the entire surface of the Earth is not realistic when the region of interest (e.g. the continental area) covers only part of it (see also section 2.3). Since extreme distortion values mostly occur along the extremities of the map, far away from the region of interest, it is logical in those cases not to include them in the analysis.

2.3 MEASURES FOR THE EVALUATION OF SMALL-SCALE MAP PROJECTION DISTORTION

The distortion measures, defined by (2.11), (2.12) and (2.13), allow us to evaluate angular distortion, area distortion and scale distortion for a specified map region S. These measures may be well suited for the evaluation of maps at a large scale, yet

Table 2.1 Mean maximum angular distortion for various equal-area projections (Behrmann, 1909).

Behrmann's results (1909)		Mean maximum angular distortion
CYLINDRICAL EQUAL-AREA PROJECTION		
Standard latitude:	0°	31°25'
	10°	29°51'
	20°	28°05'
	30°	27°06'
	40°	29°46'
	50°	38°40'
	60°	55°01'
OTHER EQUAL-AREA PROJECTIONS		
Eckert IV		27°34'
Mollweide		32°07'
Eckert VI		32°19'
Hammer–Aitoff		37°34'
Eckert II		38°18'
Sinusoidal		38°40'
Lambert's azimuthal equal-area		49°40'
Collignon		55°23'

for world maps, which are characterised by a wider range of distortion values, the squares in equations (2.12) and (2.13) give too much weight to extreme distortion. Consider, for example, the cylindrical equidistant projection with two standard parallels. Calculating the principal scale factors for each 2.5° of latitude, and minimising the overall distortion using (2.13), leads to an optimal standard latitude $\phi_0 = 69.4°$. The location of the standard parallels at this high latitude is due to the overrating of high distortion values, and produces a graticule with excessive compression of E–W distances in the lower latitudes (figure 2.2). Francula (1971), and Grafarend and Niermann (1984), using the same criterion, obtained a standard parallel at 61.7° for the mapping of areas between –85° and 85° in latitude. While the position of the standard parallels is drastically changed, when excluding the upper five degrees in latitude from the calculation, Francula (1980) admits that the resulting map does not have a favourable error distribution, the standard latitude being still too high.

In the present study, it was decided to use linear instead of quadratic measures of area and scale distortion. To prevent enlargements and reductions from compensating each other, once the squares in the expressions for overall distortion are dropped, distortion values smaller than one were substituted by their reciprocal value. While it would have been equally possible to work with absolute instead of reciprocal values, the use of the reciprocal value for scale factors smaller than one guarantees that enlargements and reductions by the same factor have equal weight in the summation. Taking into account these modifications, and assuming that distortion is calculated for a regular grid of points that is defined by a specified interval in latitude $\Delta\phi$ and in longitude $\Delta\lambda$, the integral measures for the

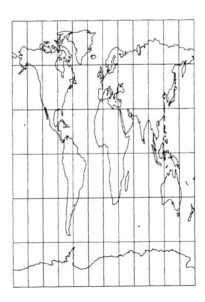

Figure 2.2 Minimum-error cylindrical equidistant projection obtained by applying Airy's least-squares criterion.

evaluation of angular, area and scale distortion that have been used in this study are the following:

$$E_{an} = \frac{1}{S} \sum_{i=1}^{k} 2 \arcsin\left(\frac{a_i - b_i}{a_i + b_i}\right) P_i \cos \phi_i \qquad (2.16)$$

$$E_{ar} = \frac{1}{S} \sum_{i=1}^{k} \left[(a_i b_i)^p - 1\right] P_i \cos \phi_i \qquad (2.17)$$

$$E_{ab} = \frac{1}{S} \sum_{i=1}^{k} \left(\frac{a_i^q + b_i^r}{2} - 1\right) P_i \cos \phi_i \qquad (2.18)$$

where k is the number of grid points and

$$S = \sum_{i=1}^{k} P_i \cos \phi_i \qquad (2.19)$$

is the sum of the area weight factors for all grid points.

Corresponding to the modifications described above, the coefficients p, q and r are defined as follows

$$p \begin{cases} = 1 & a_i b_i \geq 1 \\ = -1 & a_i b_i < 1 \end{cases}$$

$$q \begin{cases} = 1 & a_i \geq 1 \\ = -1 & a_i < 1 \end{cases}$$

$$r \begin{cases} = 1 & b_i \geq 1 \\ = -1 & b_i < 1 \end{cases}$$

As can be seen, distortion is averaged over all grid points and weighted with respect to the area surrounding each point. The factor P_i indicates whether a grid point i is located inside ($P_i = 1$) or outside ($P_i = 0$) the region of interest, allowing the integration to be restricted to this area. The *areas of major interest* concept was introduced by Robinson (1951), in a paper that propagates the use of integral measures of distortion for the evaluation of world map projections. Since most thematic world maps only represent distributions on the continental surface, Robinson argues that it is more realistic to restrict the integration of distortion to these areas. As an example, Robinson compares the mean angular and mean area distortion of the continental area for Miller's two cylindrical projections (see also section 2.5.1). He also compares the angular distortion for two interrupted equal-area graticules (Goode's Homolosine projection and Finch's interrupted Aitoff projection). Limiting the calculation of distortion to an area of interest not only allows a more realistic characterisation of overall distortion. It also creates the possibility to experiment with different aspects of the same projection, and to define optimal aspects for selected areas (see section 2.6). Projection parameters, such as the latitude of the standard parallel(s), can be adjusted to minimise distortion for any region.

Minimisation of (2.18) for the cylindrical equidistant projection, integrating distortion over the total area of the map, leads to an optimal standard latitude $\phi_0 =$ 37.5° (figure 2.3), a much more realistic value than the one obtained by minimising Airy's least-squares criterion (figure 2.2). If the calculation of distortion is restricted to the continental area, the optimal choice of the standard parallel is at 43°. This upward shift is caused by the high concentration of continental masses in the temperate zone of the Northern Hemisphere.

2.4 MEASURING OVERALL DISTORTION FOR WELL-KNOWN WORLD MAP PROJECTIONS

What follows is an attempt to evaluate existing projections for world maps, using the integral distortion measures that have been defined in the previous section. The purpose of a quantitative evaluation of world map projections is threefold. First of all, of course, it should facilitate the process of selecting a world map projection for a particular purpose. Going through the history of map projection, it seems that for most map projection designers the development of new projections for world maps was nothing more than a challenging mathematical exercise. In other words, most projections for the mapping of the world that we know today were not brought into being in reply to a specific map use. No matter how intruiging these projections may be from a mathematical point of view, they are not necessarily a suitable reference for a world map from a cartographic perspective. It is therefore important that intelligent choices are made. Most essential to realise is that there exists no such thing as *the* best projection, not for a world map, nor for the representation of any other area. One can only speak of a projection that optimally satisfies an entire set of requirements in accordance with the purpose of the map (Bugayevskiy and Snyder, 1995, p. 193). As will be shown later in this study, projection requirements may refer to the shape of the graticule (nature of meridians and parallels, level of symmetry, ...), as well as to special distortion properties (preservation of correct area, equidistancy from one point to all others, ...). A quantitative analysis of distortion may help the map maker to decide which one of the projections that satisfy all requirements will be the most suited.

Quantifying distortion for a large number of world map projections may not only support the selection process. A complete documentation of distortion characteristics (mean distortion and spatial distribution of distortion values) for a large set of map projections also provides interesting material to study the impact of graticule geometry on map projection distortion in a systematic way. As will be shown, imposing constraints on the geometry of the graticule, as dictated by the purpose of the mapping, may have a positive or negative impact on the distortion characteristics of the projection. While the definition of constraints reduces the flexibility of the graticule, some combinations of geometric properties will produce

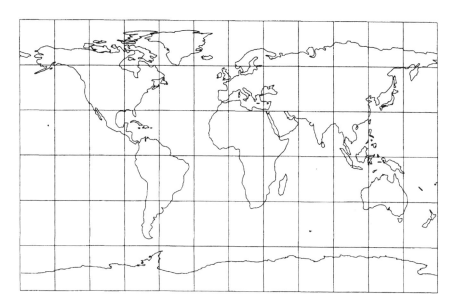

Figure 2.3 Minimum-error cylindrical equidistant projection obtained by minimising the mean linear scale distortion for the entire globe.

projections with well-balanced patterns of distortion. A proper understanding of the relationship between the geometric properties of the graticule and the distortion characteristics of the projection is important for the development of new projections, and for the optimum transformation of existing ones (see chapter 5).

Finally, it must be stressed that the results of a quantitative analysis of distortion have to be interpreted with caution. Setting up a suitability ranking of map projections based on one global distortion value may be tempting, but the results that are obtained will depend on the distortion criterion that is used. As has been explained before, the distortion measures that have been applied in this study were especially defined to avoid some of the disadvantages that are typical of more commonly used distortion measures. This, however, does not imply that they should be considered as ideal measures for the evaluation of overall distortion on small-scale maps. One of the aims of this study was also to find out if the distortion measures that have been proposed in section 2.3 produce results that agree well with the way distortion on small-scale maps is perceived, and thus, if they can actually be used for the ranking of map projections, and for the optimisation of map projection parameter values.

2.4.1 A directory of world map projections

Drawing up an inventory of all graticules ever developed for world mapping is hardly possible. The analysis was therefore restricted to those projections that are frequently quoted in the cartographic literature of the last century. The result is a directory of 68 map projections that has been published as a practical reference work (Canters and Decleir, 1989). Projections in the directory are grouped according to Maling's seven-times-three subdivision, described in section 1.4.4. Maling's classification scheme is based on the functional relationship between the coordinates on the globe and the coordinates in the plane, and is therefore directly related to the visual appearance and the distortion characteristics of the projection. Table 2.2 lists all projections that have been considered in the analysis. As can be seen, only four of Maling's seven classes were actually used. Since the introduction of a pole line has a strong impact on the distribution of distortion, an additional distinction was made between projections with, and projections without a pole line.

The information given with each projection in the directory includes the following items:

1. The name of the projection in descriptive terms. This allows one to situate the projection in the classification scheme.
2. Possible alternative names that are commonly used.
3. The name of the author and – if possible – reference to the original paper.
4. A short description of the visual appearance, the distortion characteristics and the history of the projection.
5. The transformation formulas.
6. Mean distortion values obtained by numerical integration of the angular distortion, the area distortion and the distortion of scale over the total area of the globe (E_{an}, E_{ar}, E_{ab}), as well as over the continental area only (E_{anc}, E_{arc}, E_{abc}).
7. One or two maps showing the projection's graticule, the representation of the continents, as well as lines of constant area scale and/or constant maximum angular distortion.

The six mean distortion values E_{an}, E_{ar}, E_{ab}, E_{anc}, E_{arc}, E_{abc} were obtained by applying (2.16), (2.17) and (2.18). Distortion was calculated for an interval in latitude and longitude between grid points of 2.5°. We will only concentrate on the main conclusions that can be drawn from the comparative analysis of distortion. For more information about the definition, the properties, and the distortion characteristics of the projections that were included in the analysis, the reader is referred to Canters and Decleir (1989).

2.4.2   Equal-area projections versus conformal projections

Table 2.3 lists the six mean distortion values for each projection in the directory. Plotting angular versus area distortion (figure 2.4) shows that equal-area and conformal projections can be considered as two extremes. Equal-area projections are generally characterised by a large angular distortion while conformal projections usually show a considerable distortion of area. This explains why numerous studies in the past have always valuated equal-area projections by their angular distortion (e.g. Hammer, 1889b; Bludau, 1891; Behrmann, 1909; Robinson, 1951). Values plotted on figure 2.4 have been obtained by integrating distortion over continental

Table 2.2 List of all entries in the Directory of World Map Projections (Canters and Decleir, 1989).

	POLYCONIC		PSEUDOCYLINDRICAL		PSEUDOCONICAL	CYLINDRICAL
	Pole=point	Pole=line	Pole=point	Pole=line	Pole=point	Pole=line
Decreasing parallel spacing	Hammer–Aitoff/ Nordic/Briesemeister Eckert–Greifendorff	Hammer–Wagner	Mollweide/Atlantis Putnins P2 Boggs Craster = Putnins P4 Adams Kavrayskiy V	Eckert IV Wagner IV = Putnins P2'/ Werenskiold III Wagner V Putnins P4/Werenskiold I Flat-polar parabolic Flat-polar quartic Flat-polar sinusoidal Eckert VI Wagner I = Kavrayskiy VI/ Werenskiold II Wagner II Nell–Hammer Eckert II Robinson		Cylindrical equal-area: Lambert Behrmann Peters Pavlov
Equally spaced parallels	Aitoff	Aitoff–Wagner Winkel–Tripel	Apianus II/Arago Putnins P1 Putnins P3 Sinusoidal	Ortelius Eckert III Wagner VI/Putnins P1' Kavrayskiy VII Putnins P3' Winkel II Eckert V Wagner III Winkel I Eckert I	Bonne Werner's cordiform projection	Cylindrical equidistant: Plate Carrée Equirectangular
Increasing parallel spacing	Lambert van der Grinten I			Ginzburg VIII		Mercator Miller I/Miller II Cylindrical stereographic: Braun BSAM Gall Urmayev III

Table 2.3 **(left)** Mean distortion values for each entry in the Directory of World Map Projections (Canters and Decleir, 1989).

		POLYCONIC	E_{ar}	E_{arc}	E_{an}	E_{anc}	E_{ab}	E_{abc}
		Hammer–Aitoff	0.00	0.00	35.7	33.6	0.43	0.41
		Nordic	0.00	0.00	35.7	23.7	0.43	0.26
		Briesemeister	0.00	0.00	35.8	24.0	0.47	0.26
	Decreasing spacing of parallels	Eckert–Greifendorff	0.00	0.00	35.5	35.6	0.45	0.46
Pole=point		Aitoff	0.23	0.19	30.2	28.9	0.36	0.34
	Equal spacing of parallels							
	Increasing spacing of parallels	Lambert	1.38	1.73	0.0	0.0	0.49	0.59
		van der Grinten I	1.87	1.87	7.7	7.3	0.67	0.67
		POLYCONIC	E_{ar}	E_{arc}	E_{an}	E_{anc}	E_{ab}	E_{abc}
		Hammer–Wagner	0.00	0.00	30.7	30.7	0.36	0.38
	Decreasing spacing of parallels							
Pole=line		Aitoff–Wagner	0.33	0.42	21.2	22.0	0.26	0.29
		Winkel–Tripel	0.17	0.25	23.4	22.7	0.25	0.26
	Equal spacing of parallels							
	Increasing spacing of parallels							

area. As can be verified from table 2.3 similar results are obtained by integration over the total area of the globe. Detailed examination of the distortion patterns of the projections that are included in the directory indicates that both conformal and equal-area projections always show extreme distortion in some part of the map (mostly near the edges), no matter what class of projections is considered. Hence, neither equal-area nor conformal projections are to be recommended for world maps, unless of course one of both properties is dictated by the purpose of the mapping.

Table 2.3 (right) Mean distortion values for each entry in the Directory of World Map Projections (Canters and Decleir, 1989).

PSEUDOCYLINDRICAL	E_{ar}	E_{arc}	E_{an}	E_{anc}	E_{ab}	E_{abc}	PSEUDOCONICAL	E_{ar}	E_{arc}	E_{an}	E_{anc}	E_{ab}	E_{abc}
Mollweide	0.00	0.00	32.3	33.7	0.39	0.42							
Atlantis	0.00	0.00	32.3	22.0	0.39	0.23							
Putnins P2	0.00	0.00	34.9	35.5	0.50	0.55							
Boggs	0.00	0.00	35.0	35.1	0.44	0.44							
Craster = Putnins P4	0.00	0.00	37.0	36.6	0.48	0.47							
Adams	0.00	0.00	36.0	36.4	0.47	0.48							
Kavrayskiy V	0.00	0.00	30.5	32.9	0.38	0.43							
Apianus II	0.23	0.31	24.2	25.5	0.30	0.33	Bonne	0.00	0.00	43.2	25.8	0.60	0.32
Arago	0.21	0.25	24.2	25.5	0.31	0.34	Werner	0.00	0.00	45.7	27.8	0.64	0.34
Putnins P1	0.10	0.11	30.8	31.4	0.39	0.40							
Putnins P3	0.04	0.05	35.4	35.2	0.45	0.45							
Sinusoidal	0.00	0.00	39.0	38.0	0.51	0.49							

PSEUDOCYLINDRICAL	E_{ar}	E_{arc}	E_{an}	E_{anc}	E_{ab}	E_{abc}	CYLINDRICAL	E_{ar}	E_{arc}	E_{an}	E_{anc}	E_{ab}	E_{abc}
Eckert IV	0.00	0.00	28.7	31.4	0.35	0.43	Cylindrical equal-area:						
Wagner IV = Putnins P2'	0.00	0.00	30.4	32.1	0.37	0.42	Lambert	0.00	0.00	30.9	37.8	0.55	0.77
Werenskiold III	0.34	0.34	30.4	32.1	0.39	0.44	Behrmann	0.00	0.00	26.8	32.5	0.44	0.62
Wagner V'	0.11	0.15	25.4	27.1	0.30	0.34	Peters	0.00	0.00	33.0	36.9	0.45	0.57
Putnins P4'	0.00	0.00	31.5	32.6	0.38	0.42	Pavlov	0.31	0.43	21.5	26.8	0.33	0.45
Werenskiold I	0.31	0.31	31.5	32.6	0.40	0.43							
Flat-polar parabolic	0.00	0.00	33.7	33.9	0.41	0.42							
Flat-polar quartic	0.00	0.00	32.1	33.2	0.39	0.42							
Flat-polar sinusoidal	0.00	0.00	35.1	34.6	0.42	0.42							
Eckert VI	0.00	0.00	32.4	32.9	0.39	0.42							
Wagner I = Kavrayskiy VI	0.00	0.00	31.9	32.7	0.38	0.42							
Werenskiold II	0.30	0.30	31.9	32.7	0.40	0.43							
Wagner II	0.11	0.15	26.9	27.7	0.31	0.34							
Nell–Hammer	0.00	0.00	30.9	33.6	0.42	0.50							
Eckert II	0.00	0.00	38.2	35.3	0.46	0.45							
Robinson	0.21	0.25	21.4	23.2	0.26	0.30							
Ortelius	0.39	0.54	21.6	25.2	0.29	0.38	Cylindrical equidistant:						
Eckert III	0.36	0.43	18.2	20.8	0.28	0.33	Plate Carrée	0.55	0.77	16.8	21.0	0.27	0.38
Wagner VI	0.33	0.46	20.4	22.2	0.26	0.31	Equirectangular	0.50	0.59	20.2	21.8	0.25	0.30
Putnins P1'	0.27	0.37	20.4	22.2	0.26	0.30							
Kavrayskiy VII	0.27	0.36	19.1	20.6	0.23	0.26							
Putnins P3'	0.26	0.37	22.1	23.2	0.27	0.31							
Winkel II	0.29	0.38	18.4	20.1	0.22	0.25							
Eckert V	0.28	0.34	23.4	24.0	0.30	0.32							
Wagner III	0.29	0.40	22.6	23.5	0.28	0.31							
Winkel I	0.22	0.27	25.8	25.6	0.29	0.29							
Eckert I	0.28	0.34	30.3	26.4	0.35	0.32							
Ginzburg VIII	0.49	0.59	20.3	17.5	0.29	0.29	Mercator	3.40	5.45	0.0	0.0	0.55	0.77
							Miller I	1.23	1.79	7.6	9.9	0.38	0.52
							Miller II	0.93	1.33	10.8	13.8	0.34	0.47
							Cylindrical stereographic:						
							Braun	0.96	1.36	10.3	13.3	0.35	0.47
							BSAM	0.73	1.03	9.0	11.6	0.28	0.38
							Gall	0.70	0.89	10.6	12.5	0.29	0.36
							Urmayev III	1.84	2.61	5.4	6.3	0.51	0.67

Conformal projections were never that popular for world maps, at least not for continuous representations of the whole surface of the Earth. The azimuthal stereographic projection was very much in use for world maps in the 17th century, and appears in most of the famous Dutch atlases of that period (Decleir and Canters, 1986). However, because the world was shown in two hemispheres touching along the equator, distortion was kept within acceptable bounds (figure 2.5). Since the stereographic projection depicts all meridians and parallels as arcs of

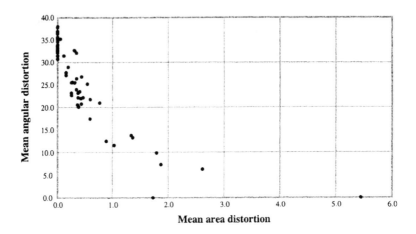

Figure 2.4 Mean maximum angular distortion versus mean area distortion for all entries in the
Directory of World Map Projections (Canters and Decleir, 1989).

circles, it is more than likely that the ease of construction has been a major
argument for the continued use of the projection, probably more important than its
conformality. Indeed, there is no real merit in the use of a conformal projection for
a world map. First there is the extreme distortion of area, impeding any assessment
of true proportions. Secondly, it is a common misunderstanding that conformal
projections preserve the shape of the continents. Conformality guarantees the
maintenance of shape only for an infinitely small area surrounding each point on
the map. The preservation of shape cannot be attained for areas of finite size.
Mercator's cylindrical conformal projection, which is the only conformal projection

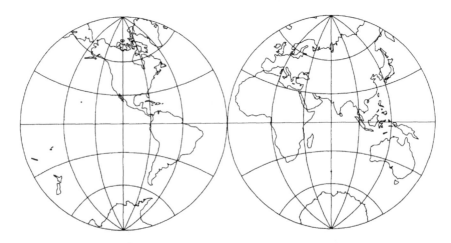

Figure 2.5 Transverse azimuthal stereographic projection in two hemispheres.

that reached a high level of popularity for the mapping of the entire world in the last century, illustrates this very clearly. While angles are preserved, the increase of area distortion with increasing latitude inevitably leads to a considerable distortion of shape in the polar areas (figure 2.6).

In spite of the persistent misuse of Mercator's conformal projection in the popular media, cartographers have always agreed that conformal projections are not suited for world maps. This has not been the case for equal-area projections. On the contrary, the equal-area property has for a long time been considered as the most important property for world maps, especially for the mapping of statistical data. Indeed, although only half of the projections listed in the directory are of the equal-area type, these projections would definitely dominate the other half if one should count their frequency of occurrence in atlases, text books, and as separately published maps. Especially the pseudocylindrical equal-area projections (e.g. sinusoidal projection, Mollweide and Eckert's graticules) have been very popular throughout the last century, in spite of their noticeable distortion of shape.

More recently, map projections with intermediate distortion characteristics (neither conformal, nor equal-area) have become increasingly popular for global mapping purposes. Because these map projections have more balanced patterns of distortion than equal-area graticules, the shape of large areas is better preserved. The Winkel–Tripel projection (figure 1.9), used in *The Times Atlas* since its first edition of 1958 (Snyder, 1993, p. 232), and the Robinson projection (figure 3.1), used by the American *National Geographic Society* from 1988 until 1998 (Garver, 1988) are two well-known examples of map projections of this kind.

If one wishes to stick to the equal-area property after all, it is obvious that the inevitable distortion of shape should be maximally reduced. Unfortunately most well known equal-area projections have not been developed with the intention of

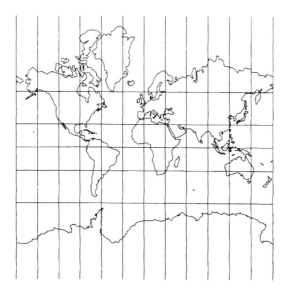

Figure 2.6 Mercator's cylindrical conformal projection.

reducing the distortion of shape. They have originated because some designer became intrigued by the mathematical problem of producing an equal-area projection, after defining a set of arbitrary constraints that control the appearance of the graticule (Robinson, 1951). Most of these projections are characterised by straight, circular, elliptical, parabolic or sinusoidal meridians, they represent the pole as a point, or they have pole lines that are exactly half or a third the length of the equator (see also section 2.5). Although they can be nicely described mathematically, the majority of these map projections has no apparent cartographic utility. Indeed, by first decreeing through formulas the nature of the meridional curves and the length of the pole line, the shape of the continents has already been partly decided. At best, a few variables in the transformation formulas still allow some minor modifications to be made. When developing new projections for world maps, one should proceed differently. Indeed, the proper representation of large areas should become an essential part of the design process, for equal-area as well as for non equal-area maps. Recent efforts in this direction will be discussed in chapter 3.

2.4.3   Mean distortion of scale as a hybrid measure of distortion

Since conformal projections are characterised by an extreme distortion of area, they can be evaluated by calculating a mean area distortion value using equation (2.17). In a similar way, a comparison of equal-area projections can be done by calculating for each projection a mean angular distortion value, applying equation (2.16). Projections that are neither conformal nor equal-area have intermediate distortion characteristics. In order to evaluate these projections, and compare their overall distortion with conformal and equal-area projections, there is a need for a single measure of distortion that quantifies the combined effect of both angular and area distortion. Since angular and area distortion both result from a scale distortion, varying from point to point and with direction, it seems logical to evaluate map projections by calculating the mean linear scale distortion, defined by equation (2.18).

A quick look at table 2.3 immediately shows that equal-area and conformal projections have the highest scale distortion. Projections that are neither conformal nor equal-area score remarkably better. This confirms what has already been said about the unsuitability of equal-area and conformal projections for world maps. Especially the polyconic and pseudocylindrical projections with small area distortion have favourable values for the mean linear scale distortion. Of all graticules that are symmetric about the equator Winkel II (figure 2.7), Kavrayskiy VII (figure 2.8), and Winkel–Tripel (figure 1.9) have the lowest value. All three projections have similar geometric characteristics. They all have parallels that are equally spaced along the central meridian, as well as a pole line half the length (Kavrayskiy VII), or less than half the length of the equator (Winkel II, Winkel–Tripel). Parallels and meridians never meet at very sharp angles. For each projection minimum distortion occurs in the middle latitudes (see figure 1.9), and not on the equator, as it is the case for most other polyconic and pseudocylindrical projections with equally spaced parallels. The foregoing suggests a strong relationship between the geometric characteristics of the graticule and the distortion

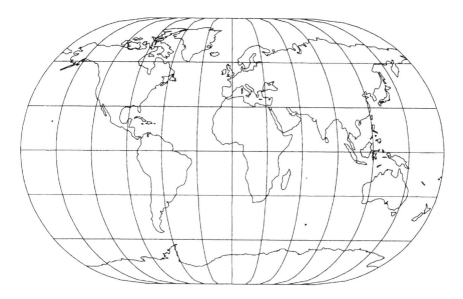

Figure 2.7 Winkel's second projection (Winkel II).

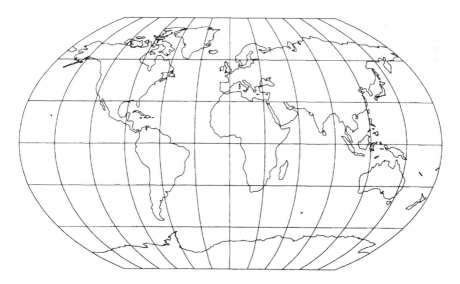

Figure 2.8 Kavrayskiy's seventh projection (Kavrayskiy VII).

patterns of a projection, as could be expected. In the next section, we will examine in detail how each of the graticule's characteristics influences the distortion properties of the projection. This will give us useful information for the development of new map projections.

2.5   INFLUENCE OF MAP PROJECTION GEOMETRY ON DISTORTION CHARACTERISTICS

2.5.1   Spacing of the parallels

As indicated in figure 2.4, the distortion characteristics of a projection seem to be related to the spacing of the parallels. Conformal projections, as well as projections with small angular distortion, are characterised by increased parallel spacing from the equator towards the poles, while most equal-area projections and projections with small area distortion have decreased parallel spacing. Graticules with equally spaced parallels occupy an intermediate position. In consequence, they have a lower mean linear scale distortion. The phenomenon can be demonstrated at best by comparing a set of map projections that differ from one another by the spacing of the parallels only. This is for example the case for all cylindrical projections that have the equator as the standard parallel.

In 1942, O.M. Miller generalised the formulas of the Mercator projection by introducing a constant C in the expression for the y-coordinate (Miller, 1942):

$$y = CR \ln \tan\left(\frac{\pi}{4} + \frac{\phi}{2C}\right) \qquad\qquad (2.20)$$

For $C=1$ the Mercator projection itself is obtained. Increasing the value of C allows for a reduction of the area distortion at the cost of a growing angular distortion (Canters and Decleir, 1989, p. 168). Miller defined two new map projections, characterised by $C=1.25$ and $C=1.5$ respectively. His formulas, however, make it possible to generate an unlimited number of cylindrical projections, all having increased parallel spacing from the equator towards the poles. The rate of increase will systematically diminish as the value of C is raised. Ultimately one obtains the cylindrical equidistant projection. Figure 2.9 shows how the mean linear scale distortion for Miller's projection decreases for progressively higher values of C. Applying (2.18), it can be shown that the mean linear scale distortion reaches its minimum (0.275) for a value of C equal to infinity, in other words, for the cylindrical equidistant projection.

Similar to Miller's transformation, a sequence of cylindrical projections with decreasing parallel spacing can be generated by introducing a constant C in the expression for the y-coordinate of the cylindrical equal-area projection:

$$y = CR \sin\frac{\phi}{C} \qquad\qquad (2.21)$$

This sequence also converges to a cylindrical equidistant projection as the value of

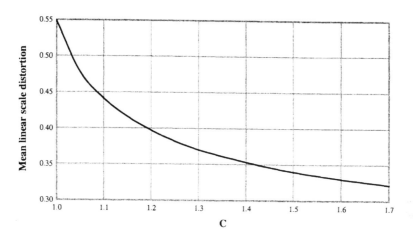

Figure 2.9 Mean linear scale distortion as a function of the parameter C in Miller's generalised Mercator projection.

C is increased. The mean linear scale distortion, as defined by equation (2.18), is exactly the same as for Miller's generalised projection, which means that also in this case the minimum value is obtained for an equal spacing of the parallels (figure 2.9).

From the foregoing one may conclude that the spacing of the parallels has a strong impact on the distortion characteristics of a projection. According to the criterion used in this study, it seems that an optimal balance between the distortion of angles and area is obtained for an equal spacing of the parallels. A comparison of the graticules included in the directory indeed shows that an equal spacing of the parallels generally has a pleasing effect on the representation of the continents. It may be interesting to note that this equal spacing of the parallels also occurs on the undistorted globe.

2.5.2   Length of the pole line

The impact of a pole line on the distortion pattern of a map projection is well known. While pointed-polar projections tend to compress the northern landmasses, cylindrical projections, showing the pole as a line with the same length as the equator, have an unacceptable E–W stretching in the higher latitudes. Obviously one may expect to obtain a more balanced distortion pattern by introducing a straight pole line of intermediate length. Hence it is not surprising that numerous projections of this type have been developed, especially for world mapping.

The earliest world maps with a straight pole line shorter than the equator belong to the group of so-called *oval* maps of the sixteenth century. One variant of the oval projection has rectilinear parallels that are equally spaced along the central meridian, curved meridians, a 2:1 ratio of the axes, and a pole line half the length of the equator. The famous world map in Abraham Ortelius' *Theatrum orbis terrarum* (1570) is probably the best-known example of the use of this type of projection (figure 2.10), yet it was not the first. Snyder (1993, p. 38) states that the same projection was already used by Battista Agnese about 30 years earlier. Although very popular at the time oval projections with flat poles were seldom used in the next centuries. The use of a pole line was reintroduced in 1890 by Nell, who developed the first *true* pseudocylindrical projection (see section 1.4.1) with poles shown as lines rather than points (Nell, 1890). In the beginning of the twentieth century, many new pseudocylindrical and polyconic projections with pole line were proposed of which several are still in use today.

A simple way of creating a projection with a pole line is by averaging the coordinates of a pointed-polar projection and a cylindrical projection. In 1906, Max Eckert defined three new pseudocylindrical projections with equally spaced parallels (Eckert I, Eckert III, Eckert V) (figure 2.11). All three are characterised by a correct ratio of the axes and a pole line half the length of the equator, just like the oval projection discussed above. The three projections only differ by the nature of the meridians. Each projection is defined as the arithmetic mean of a pseudocylindrical projection that shows the pole as a point, and the cylindrical equidistant projection with one standard parallel (*Plate Carrée*). The axes are scaled to make the total area of the planisphere equal to the area of the globe. This allowed Eckert to derive also three equal-area versions of these projections by properly adjusting the spacing of the parallels (Eckert II, Eckert IV, and Eckert VI respectively) (Eckert, 1906).

Eckert's first projection has rectilinear meridians. It is the simplest of all but it has no real practical value. Since the meridians are straight lines, they meet at an angle on the equator. This discontinuity gives the projection a very unpleasant appearance. Eckert III (elliptical meridians) and Eckert V (sinusoidal meridians) have more balanced distortion patterns. A comparison with their parent projections (Apianus' elliptical projection and Sanson's sinusoidal projection respectively) shows how the introduction of a pole line reduces the angular distortion, while the

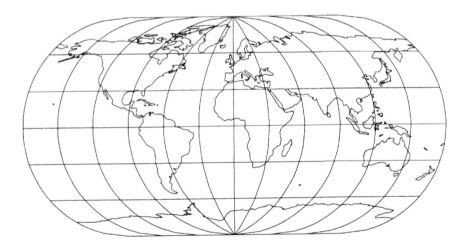

Figure 2.10 Ortelius' oval projection with equally spaced parallels, circular meridians, a pole line half the length of the equator, and a 2:1 ratio of the axes.

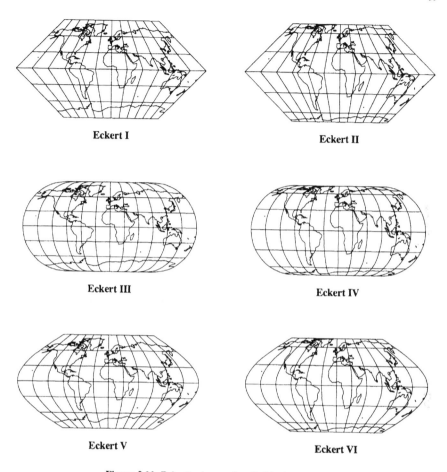

Figure 2.11 Eckert's six pseudocylindrical projections.

area distortion is increased (table 2.4). The overall improvement that can be achieved by introducing a pole line depends on the initial shortcomings of the parent projection. While Sanson's sinusoidal projection (figure 2.14c) is characterised by a substantial angular distortion that becomes especially

Table 2.4 Mean distortion values for Eckert III and Eckert V, and for their pointed-polar parent projections.

	E_{ar}	E_{arc}	E_{an}	E_{anc}	E_{ab}	E_{abc}
Eckert III	0.36	0.43	18.2	20.8	0.28	0.33
Eckert V	0.28	0.34	23.4	24.0	0.30	0.32
Apianus II	0.23	0.31	24.2	25.5	0.30	0.33
Sinusoidal	0.00	0.00	39.0	38.0	0.51	0.49

conspicuous in the high latitudes, Eckert's fifth projection gives a far better portrayal of the continental configuration. This is also reflected in the mean linear scale distortion for both projections. For Apianus' elliptical projection, which has no excessive shearing in the high latitudes – the projection is identical in appearance to Arago's projection (figure 2.13a) – the balance between angular and area distortion is not clearly improved by the introduction of a pole line. The mean linear scale distortion values for Eckert III are almost identical to those obtained for Apianus' elliptical projection. Due to the moderate curvature of the meridians, the variation of area scale on Eckert III is quite high (Canters and Decleir, 1989, p. 85).

The equal-area versions of Eckert's first three projections differ from their parent graticules by the spacing of the parallels only. As can be seen in figure 2.11, the equal-area constraint imposes a decrease in the parallel spacing from the equator towards the poles, which leads to a disturbing N–S stretching of the equatorial areas. This confirms the earlier remark that all equal-area projections introduce excessive shear somewhere. While pointed-polar equal-area projections show a high-latitude crowding, flat-polar graticules introduce a N–S stretching of the equatorial areas that is combined with a stretching in the opposite direction in the higher latitudes.

Eckert's work inspired many authors to develop new projections that represent the pole as a line. O. Winkel generalised Eckert's construction principle by averaging pointed-polar projections with the cylindrical equidistant projection with two standard parallels (Winkel, 1921). Starting from a pointed-polar projection with true scale along the central meridian, it allows for the development of an unlimited number of projections only differing from one another in the length of the pole line and the ratio of the axes. Winkel proposed two new pseudocylindrical projections by averaging the cylindrical equidistant projection that has the same total area as the globe (standard parallels at 50°28') with Sanson's sinusoidal projection (Winkel I), and with Apianus' elliptical projection (Winkel II). As indicated before, Winkel's second projection (figure 2.7) has the lowest mean linear scale distortion of all projections in the directory (table 2.3). Winkel's most famous projection, however, is his third, which is the arithmetic average of the cylindrical equidistant projection and the Aitoff projection (see Canters and Decleir, 1989, p. 51), and which is known as the Winkel–Tripel (figure 1.9). The projection became very popular through its use in *The Times Atlas of the World,* and is currently also used by the American *National Geographic Society.* Like on Winkel's other two projections, parallels are equidistantly spaced along the central meridian, and the pole line is less than half the length of the equator. Since the

Table 2.5 Nomenclature of Putnins' twelve pseudocylindrical projections.

	Equally spaced parallels		Decreasing parallel spacing (equal-area)	
	Pole=point	Pole=line	Pole=point	Pole=line
Elliptical meridians	P1	P1'	P2	P2'
Parabolic meridians	P3	P3'	P4	P4'
Hyperbolic meridians	P5	P5'	P6	P6'

Aitoff projection is not of the pseudocylindrical type, the parallels on the Winkel–Tripel are slightly curved.

R.V. Putnins (1934) proposed no less than twelve conic section pseudocylindrical projections (table 2.5), of which three (Putnins P1', Putnins P3',Putnins P5') are obtained by applying Eckert's averaging principle. Just like Eckert's graticules, they have a pole line half the length of the equator, and a correct ratio of the axes. Only the nature of the meridians is different. A comparison of P1' and P3' with their respective parent projections P1 and P3 (figure 2.13b, figure 2.13c) again shows that the introduction of a pole line lowers the angular distortion, while the distortion of area is increased (table 2.3). Mean linear scale distortion values, and examination of the distortion patterns of the projections, indicate that the flat-polar projections are superior to their related pointed-polar versions. Putnins also developed three pointed polar equal-area projections, and three equal-area versions of his flat-polar projections. Apart from a different scale factor, the three flat-polar projections differ from their parent graticules by the spacing of the parallels only. Eight of Putnins' twelve projections (those that have elliptical or parabolic meridians) are included in the directory. For more information about these projections, see Canters and Decleir (1989, pp. 77–9, p. 87, p. 91, p. 109, p. 113, p. 121, p. 125).

In 1949, F.W. McBryde and P. Thomas proposed a new method for the development of pseudocylindrical equal-area projections with a pole line of given length. By applying a general transformation that preserves the equal-area property of the pointed-polar parent projection, they developed a number of flat-polar projections that all have a pole line less than half the length of the equator (McBryde and Thomas, 1949). It was the intention of the authors to reduce the N–S stretching in the lower latitudes, as well as the E–W stretching in the higher latitudes, through the shortening of the pole line. Three of their projections are included in the directory: the flat-polar parabolic, the flat-polar quartic, and the flat-polar sinusoidal projection (table 2.2). Each has a pole line one-third the length of the equator. For all three projections the scale variation along the central meridian is considerably less than on Eckert's and Putnins' graticules (Canters and Decleir, 1989, pp. 127–31). The flat-polar quartic projection (figure 2.12), which is the most popular of the three, is to be preferred since it gives a better representation of the polar regions than the flat-polar parabolic and the flat-polar sinusoidal. This is also reflected in its slightly lower mean angular distortion values (table 2.3). Still the equal-area property leaves its inevitable trace. The shortening of the pole line leads to an increased angular distortion close to the bounding meridian, which indicates that a perfect balance between pointed-polar and flat-polar projections is difficult to obtain.

Well before McBryde and Thomas published their work, Wagner had already proposed a similar, but more general method for the development of flat-polar projections with a pole line of controllable length (Wagner, 1932). By introducing additional parameters in the transformation formulas of existing map projections, Wagner created generalised map projections that include their parent projection as a special case. Wagner's method, which is known in the German literature as *Das Umbeziffern*, and which was formally described by Siemon in 1936, is very flexible and permits extensive modification of the distortion pattern of a projection.

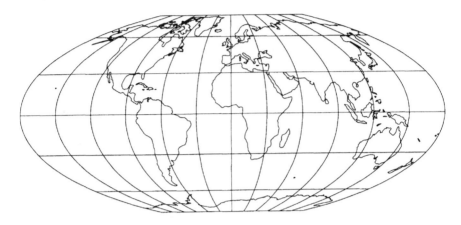

Figure 2.12 McBryde and Thomas' flat-polar quartic projection.

The use of the technique is discussed in detail in the fourth chapter of this study (see section 4.1.2). As will be shown, the graticule of the parent projection can be transformed without altering the equal spacing of the parallels or without changing the pattern of area distortion. Especially the last type of transformation is very interesting since it offers the possibility to derive new equal-area graticules from existing ones. Examples of flat-polar map projections that are obtained by applying both variants of Wagner's technique, and that have been proposed by Wagner himself (Wagner, 1949), are included in the directory and will be discussed in chapter 4. Those with elliptical meridians (Wagner IV, Wagner VI) (figure 4.2) are generalised versions of projections that were developed by Putnins fifteen years earlier (Putnins P2', Putnins P1') (see above). Those with sinusoidal meridians (Wagner I, Wagner III) (figure 4.3) highly resemble Eckert's sinusoidal projections (Eckert VI, Eckert V).

Being aware of the constraints imposed by the equal-area property, Wagner (1949) also described a third variant of his technique, which allows the introduction of an area distortion that increases with latitude (see section 4.1.2). It concerns a generalisation of the transformation that maintains the original pattern of area distortion. Starting from an equal-area graticule, the technique allows the development of projections with an area distortion that increases from 1.0 on the equator to a specified value on a chosen parallel. Wagner himself proposed two pseudocylindrical projections of this kind by specifying an area distortion of 20% at 60° latitude (Wagner II, Wagner V) (figure 4.4). This implies that almost all populated areas of the world are represented with less than 20% area distortion. Both projections show considerably less scale variation along the central meridian than pseudocylindrical equal-area projections with a pole line. They have an area distortion that is much lower than for pseudocylindrical projections with equally spaced parallels and pole line, at the cost of a higher angular distortion (table 2.3). Still, the mean linear scale distortion remains the lowest for projections with equally spaced parallels, which is once again an indication of the good balance between angular and area distortion for the latter (see also section 2.5.1).

The use of Wagner's transformation method is not restricted to the pseudocylindrical class of projections. It can also be applied to graticules with curved parallels. The well-known Aitoff transformation, proposed by David Aitoff in 1889, and applied by Ernst Hammer in 1892 to construct his famous Hammer–Aitoff projection (Hammer, 1892), is in fact a simple case of Wagner's transformation. Applying the transformation in its generalised form, Wagner (1949) proposed two polyconic projections with curved pole line (Aitoff–Wagner, Hammer–Wagner), both avoiding the compression in the high latitudes so characteristic of the original pointed-polar graticules (Aitoff, Hammer–Aitoff) (see section 4.1.2). The Aitoff–Wagner projection, with its slightly curved, equally spaced parallels, and its pole line shorter than half the length of the equator (figure 4.6), strongly resembles the Winkel–Tripel projection. The Hammer–Wagner projection (figure 4.8), which is equal-area, has lower mean distortion values for the continental area than all pseudocylindrical equal-area projections with pole line that have been considered in this study. Nevertheless the strong, unpleasant variation in scale along the central meridian remains. For that reason Wagner (1949) also proposed a polyconic projection with curved pole line and prescribed area distortion (figure 4.9), comparable with Wagner II and Wagner V (see above).

2.5.3 Nature of the meridians

It is understood that the nature of the meridians highly influences the appearance of the projection. Table 2.6 lists four pointed-polar pseudocylindrical projections with equally spaced parallels, a 2:1 ratio of the axes and a total area that equals the area of the generating globe. The four projections, which are also shown in figure 2.13 and figure 2.14c, have been ordered according to the steepness of the meridians, which is the only variable in their construction. As can be seen, the nature of the meridians has a clear impact on the distortion characteristics. As the meridians converge sharper (from full semi-ellipses to full cosinusoids) the angular distortion increases in favour of the representation of area. The scale factor along both axes of the projection grows to compensate for the increasing steepness of the meridians, so that the total area of the graticule remains equal to the area of the generating globe. Mean linear scale distortion values show how the overall balance of distortion improves as the meridians meet at a less sharp angle (table 2.3). This also holds for other types of projections, although it must be pointed out that the influence of meridional steepness on distortion characteristics is less for projections

Table 2.6 Distortion characteristics for pointed-polar pseudocylindrical projections with different curvature of the meridians

	Nature of the meridians	Mean area distortion	Mean maximum angular distortion	Scale factor along the axes
Arago	Full ellipses	0.21	24.2	0.90
Putnins P1	Portions of ellipses	0.10	30.8	0.95
Putnins P3	Parabolas	0.04	35.4	0.98
Sinusoidal	Full cosinusoids	0.00	39.0	1.00

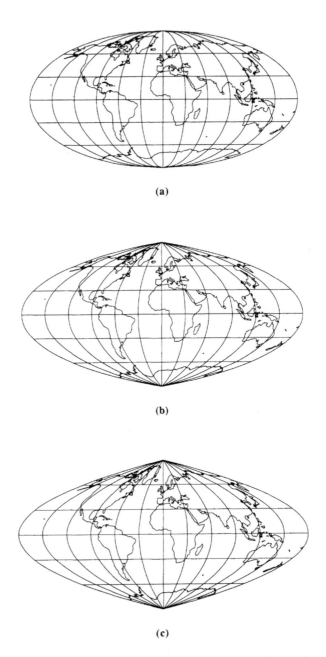

(a)

(b)

(c)

Figure 2.13 Three pointed-polar pseudocylindrical projections with equally spaced parallels, a 2:1 ratio of the axes, and a total area that equals the area of the generating globe: Arago's projection (a), Putnins P1 (b), and Putnins P3 (c) (all graticules are on the same scale).

with a pole line than for projections that represent the pole as a point. This, of course, is not surprising. The main reason for the creation of a pole line is exactly to avoid the high-latitude crowding, which is caused by a sharp convergence of the meridians in the first place.

For equal-area graticules, the nature of the meridians is closely tied to the spacing of the parallels. Going from Mollweide's projection (figure 2.14a), which encloses the world within an ellipse, to Sanson's sinusoidal projection (figure 2.14c) with its sharply converging meridians, the E–W compression in the high latitudes increases. Yet the scale variation along the central meridian, which explains the relative N–S stretching of the equatorial areas on Mollweide's graticule, becomes less extreme. Sanson's projection shows no scale variation along both axes. All parallels are equally spaced along the central meridian. Being aware of the deficiencies of both the Mollweide and the sinusoidal projection, several authors developed projections with intermediate characteristics. Most remarkable is W. Boggs' *eumorphic* projection (figure 2.14b). It was developed by averaging the *y*-coordinates of Sanson's and Mollweide's graticules, and maintaining the equal-area property through the adjustment of the *x*-coordinate (Boggs, 1929). With the term *eumorphism*, Boggs is one of the first to refer to a good representation of large shapes, which is, as we will see in the next chapter, a major concern in the development of world maps today.

When examining equal-area projections with a pole line, it is again clear that the nature of the meridians has a much smaller impact on distortion characteristics than it is the case for pointed-polar equal-area projections. As was demonstrated in the previous section, the scale variation along the central meridian on flat-polar equal-area projections is mainly determined by the length of the pole line. Meridional steepness mostly influences the amount of angular distortion near the edges of the map. As the meridians become "rounder", angular distortion is somewhat decreased.

2.5.4 Ratio of the axes

Although on the globe the equator is twice as long as the central meridian, the results of the distortion analysis show that the projections that have the lowest mean scale distortion (Winkel II, Kavrayskiy VII, Winkel–Tripel) all have a central meridian that is more than half the length of the equator. A closer examination of these projections helps to clarify things. On all three projections (figure 2.7, figure 2.8, figure 1.9) the N–S stretching of the continents immediately attracts attention. Compared with similar graticules that have a 2:1 ratio of the axes (Eckert III, Wagner VI), this relative lengthening of the central meridian compensates for the E–W stretching in the high latitudes. Similar to the introduction of two standard parallels on a cylindrical projection, the adjustment of the ratio of the axes causes a redistribution of distortion throughout the whole map area. Scale distortion reaches a minimum in the middle latitudes instead of on the equator, which leads to less extreme distortion near the edges of the map. This explains why more favourable overall distortion values are obtained than for map projections for which the ratio of the axes is correct.

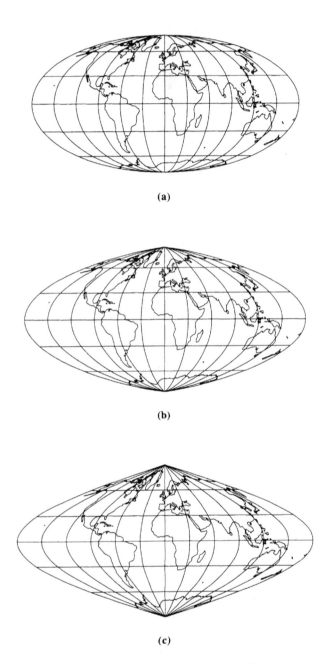

(a)

(b)

(c)

Figure 2.14 Three pointed-polar pseudocylindrical equal-area projections: Mollweide's projection
(a), Boggs' eumorphic projection (b), and Sanson's sinusoidal projection (c) (all graticules are on the
same scale).

On the other hand, one should not deviate too far from the correct ratio of the axes, as the eye proves to be very sensitive to shape distortion in the lower latitudes. Since the African continent always occupies a central position on conventional world maps, even small deviations from its original shape are easily perceived. While the N–S stretching on Kavrayskiy VII still seems acceptable (ratio of the axes 1.73), it is much more disturbing on Winkel II (ratio of the axes 1.64). This difference cannot be inferred from the mean distortion values, which again indicates that the results of a quantitative analysis of map projection distortion should be interpreted with caution.

2.6 DIFFERENT ASPECTS OF A MAP PROJECTION

One of the most effective ways to reduce distortion is by centring the distortion pattern of the projection on another part of the Earth's surface, in order to place the region of interest in the least distorted area of the map (mostly close to the centre of the projection). Such a repositioning is denoted a *change of aspect*, and somewhat corresponds with looking at the Earth from another direction (see also section 1.4.1).

To develop a map projection, a reference system has to be established on the surface of the sphere. Geographical longitude λ and latitude ϕ, which are mostly used to describe the position on the globe, are defined with respect to a mathematical pole P, which coincides with the geographical North Pole, and a corresponding reference great circle Q, which is called the equator (figure 2.15). In general, however, the position on the globe can equally well be defined with respect to a reference system determined by an arbitrary great circle Q' that is oblique to the equator. The associated mathematical pole T will then no longer coincide with one of the geographical poles. Great circles passing through the mathematical pole are in that case referred to as *meta-meridians*, while small circles perpendicular to the polar axis are called *meta-parallels*. Once such a reference system has been established, each point on the globe's surface can be described by its *meta-longitude* λ' and *meta-latitude* ϕ' (figure 2.15).

Map projections can be defined with respect to any spherical reference system of the type described above by simply substituting the meta-longitude and the meta-latitude for the longitude and the latitude in the projection's transformation formulas:

$$x = f_1(\lambda',\phi')$$
$$y = f_2(\lambda',\phi') \tag{2.22}$$

To be able to calculate the position on the map for a point M with given latitude and longitude, the calculation of (2.22) has to be preceded by a coordinate transformation that derives the meta-longitude λ' and meta-latitude ϕ' of the point from its geographical coordinates λ and ϕ:

$$\lambda' = f_3(\lambda,\phi)$$
$$\phi' = f_4(\lambda,\phi) \tag{2.23}$$

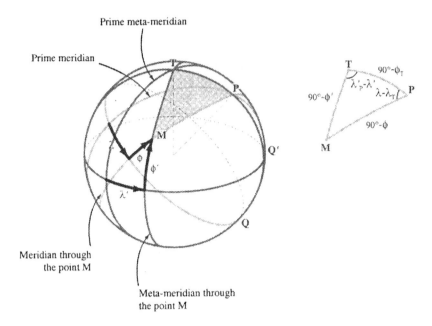

Figure 2.15 Relation between meta-coordinates and geographical coordinates.

The transformation is accomplished by solving the spherical triangle *PTM* (figure 2.15):

$$\sin\phi' = \sin\phi\sin\phi_T + \cos\phi\cos\phi_T\cos(\lambda - \lambda_T) \qquad (2.24)$$

$$\sin(\lambda'_p - \lambda') = \frac{\sin(\lambda - \lambda_T)\cos\phi}{\cos\phi'} \qquad (2.25)$$

$$\cos(\lambda'_p - \lambda') = \frac{\sin\phi\cos\phi_T - \cos\phi\sin\phi_T\cos(\lambda - \lambda_T)}{\cos\phi'} \qquad (2.26)$$

where λ_T and ϕ_T are the geographical coordinates of the meta-pole, and λ'_p is the meta-longitude of the geographical pole. The meta-latitude ($-90° \leq \phi' \leq 90°$) is unambiguously determined by equation (2.24). For the meta-longitude ($-180° \leq \lambda'$ $\leq 180°$) quadrant adjustment is necessary. Calculation of $\sin(\lambda'_p-\lambda')$ and $\cos(\lambda'_p-\lambda')$ from equations (2.25) and (2.26) makes identification of the quadrant possible, and yields the correct value for λ'.

    A change of aspect leaves the distortion pattern of the projection unchanged, yet shifts its position with respect to the Earth's surface. This will alter the appearance of the graticule and the representation of the continents. Indeed, the meta-pole will now take the position of the geographical pole, while the meta-equator will take the position of the equator in the projection. The meta-parallel of

30° will take the place of the parallel of 30°, and so on. The less familiar representation of the original graticule, and the repositioning of the continents caused by a change of aspect may confuse the untrained map user (see below). Yet it allows any region on the globe to be put in the least distorted area of the map. Mean distortion values obtained by integration of distortion over the whole surface of the Earth will not be changed. Integration of distortion over the continental area, however, will yield different results depending on the location of the projection centre, and on the orientation of the axes of symmetry of the distortion pattern.

Traditionally three aspects of a map projection are distinguished: the *normal* (or *direct*) aspect, the *transverse* aspect, and the *oblique* aspect. In the normal aspect of a projection, the geographical graticule and the *meta-graticule* coincide, and there is a direct relationship between the appearance of the meridians and the parallels and the distortion pattern of the projection. In the transverse aspect graticule and meta-graticule are at right angles. In the oblique aspect they are at any arbitrary angle. A change of aspect does not affect the representation of the meta-graticule. For every aspect of the projection, the meta-graticule coincides with the representation of the system of meridians and parallels in the normal aspect. Figure 2.16 gives an example of a normal, oblique, and transverse aspect of the cylindrical equidistant projection. As can be seen, for the normal aspect the geographical poles coincide with the poles of the projection (meta-poles) (figure 2.16a). For the transverse aspect, they are located on the equator of the projection (meta-equator) (figure 2.16c). In the oblique aspect that is shown in figure 2.16b the position of the meta-pole has been chosen at 45°N, 180°E. Hence the North Pole is situated at a meta-latitude of 45°, and a meta-longitude of 0°. For the South Pole, the meta-coordinates are –45°, 180°.

The recognition of three different aspects is satisfactory for the conic group of projections (including the azimuthal, conical and cylindrical class), yet it fails to describe all possible appearances of the graticule for projections that have a lower level of symmetry. Most projections for world maps belong to the polyconic or pseudocylindrical class and have only two axes of symmetry, i.e. a straight meta-equator, and a straight prime meta-meridian. For these projections no less than seven different aspects can be distinguished, depending on the location of the geographical pole on a trirectangular spherical triangle defined by the three basic great circles of the meta-graticule (figure 2.17) (Wray, 1974). Figure 2.18 illustrates all seven aspects for the Mollweide projection. The first six aspects are obtained by moving the geographical pole along the perimeter of the spherical triangle, shown in figure 2.17. In the *normal* aspect, the polar axis coincides with the Z-axis. If the polar axis is rotated over 90° in the XZ-plane the *first transverse* aspect is obtained, if it is rotated over 90° in the YZ-plane one obtains the *second transverse* aspect. The three together form the *orthogonal* aspects, with the polar axis coinciding with one of the three coordinate axes X, Y, Z. In case the polar axis lays in the XZ-, XY- or YZ-plane, the aspect is called *simple oblique, transverse oblique,* and *skew* respectively. These three aspects are also referred to as the *diagonal* aspects. If the polar axis is oblique to all three coordinate axes, the *plagal* aspect is obtained.

Equations (2.24)–(2.26) are for the general, plagal case. For the other aspects equations will simplify, depending on the latitude of the meta-pole and the meta-

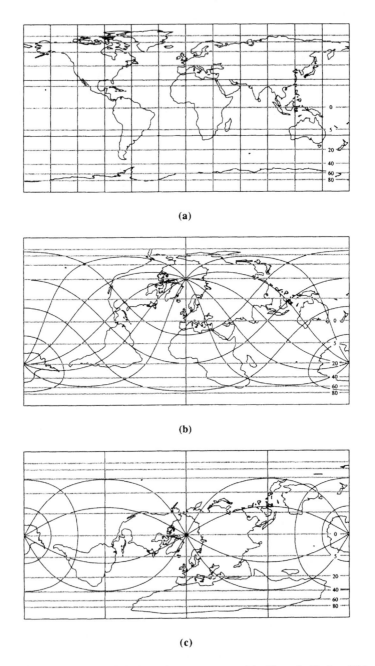

(a)

(b)

(c)

Figure 2.16 Normal (a), oblique (b), and transverse aspect (c) of the cylindrical equidistant projection, with lines of constant maximum angular distortion.

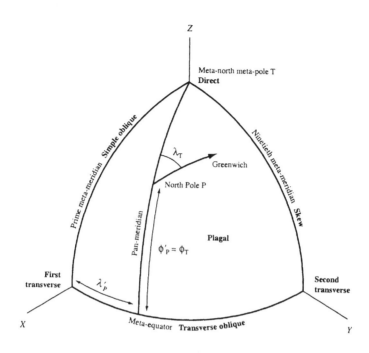

Figure 2.17 Trirectangular triangle used by Wray to define the seven aspects of a non-conic map projection (Wray, 1974, p. 8; reprinted by permission of University of Toronto Press Incorporated).

longitude of the North Pole (table 2.7). Since for the conic group of projections all meta-meridians are alike (the conic group possesses rotational symmetry about its meta-polar axis), it follows that the first transverse, the second transverse and the transverse oblique aspects constitute one single transverse aspect. Similarly, the simple oblique, the skew oblique and the plagal aspects constitute one single oblique aspect. It should be mentioned that Wray also distinguishes four independent *equioblique sub-aspects*, one for each of the four non-orthogonal aspects. For the three diagonal aspects the equioblique sub-aspects are called *simple equioblique*, *transverse equioblique*, and *equiskew* respectively, and are obtained by letting the polar axis make an angle of 45° with each of the two coordinate axes that define the plane in which it lies. The *equiplagal* aspect is obtained by letting the polar axis make equal angles with all three axes (54°44'). Each of the four non-orthogonal aspects of the Mollweide projection that are shown in figure 2.18 corresponds with the equioblique sub-aspect.

According to Wray (1974), more than thirty indirect aspects of non-conic projections have been described in the cartographic literature of the last century. Only one of these is a skew aspect, three are plagal aspects. Thus practically all indirect aspects of projections ever proposed have a graticule that has at least one axis of symmetry. Of these graticules almost all that have been used for world maps represent the world within an ellipse, and are derived from Mollweide's pseudocylindrical or Hammer's polyconic projection, two equal-area graticules that

Table 2.7 Aspect parameter values for the seven different aspects of a non-conic map projection.

Aspect	Latitude of the meta-pole	Meta-longitude of the North Pole	Comments
Normal	90°	180°	the longitude of the meta-pole in eqs. (2.24) – (2.26) defines the position of the central meridian
First transverse	0°	0°	North Pole in the centre of the map
		180°	South Pole in the centre of the map
Second transverse	0°	90°	North Pole located on the meta-east side of the prime meta-meridian
		-90°	North Pole located on the meta-west side of the prime meta-meridian
Transverse oblique	0°	free	
		45°, 135°	transverse equioblique aspect, North Pole located on the meta-east side of the prime meta-meridian
		-45°, -135°	transverse equioblique aspect, North Pole located on the meta-west side of the prime meta-meridian
Simple oblique aspect	free	0°	North Pole located on the prime meta-meridian
		180°	South Pole located on the prime meta-meridian
	45°	0°	simple equioblique aspect, North Pole located on the prime meta-meridian
		180°	simple equioblique aspect, South Pole located on the prime meta-meridian
Skew aspect	free	90°	North Pole located on the meta-east side of the prime meta-meridian
		-90°	North Pole located on the meta-west side of the prime meta-meridian
	45°	90°	equiskew aspect, North Pole located on the meta-east side of the prime meta-meridian
		-90°	equiskew aspect, North Pole located on the meta-west side of the prime meta-meridian
Plagal aspect	free	free	
	35°16'	45°, 135°	equiplagal aspect, North Pole located on the meta-east side of the prime meta-meridian
		-45°, -135°	equiplagal aspect, North Pole located on the meta-west side of the prime meta-meridian

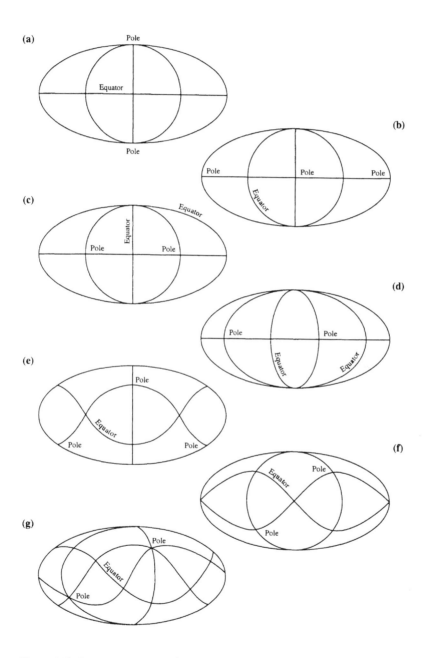

Figure 2.18 Wray's seven aspects of a non-conic map projection, shown for the Mollweide graticule: normal (a), first transverse (b), second transverse (c), transverse oblique (d), simple oblique (e), skew (f), and plagal aspect (g) (Wray, 1974, p. 17-23; reprinted by permission of University of Toronto Press Incorporated).

represent the pole as a point. One of the best known is Bartholomew's Nordic projection, a simple oblique aspect of Hammer's graticule, first used in 1950 (Snyder, 1993, p. 239), and included in many editions of *The Times Atlas of the World* (figure 2.19). Also Briesemeister's projection is relatively well known (Briesemeister, 1953). Apart from the ratio of the axes and the eastward shift of the central meridian, it is identical to the Nordic projection (figure 2.20). Two other interesting examples of the use of indirect aspects are the conformal and equal-area world ocean maps developed by Spilhaus (1942). Both maps are transverse oblique aspects (of August's conformal and Hammer's equal-area projection respectively) with a bounding meta-meridian that is entirely within land, except for the crossing of minor waters. Hence all oceans are shown without interruption. More recently, the same author presented several other examples of world ocean maps by choosing proper aspects of standard map projections. Yet instead of interrupting the graticule at the outer meta-meridian, he extended it beyond the meta-longitude of 180° to be able to show both oceans and continents in their entirety. All maps have irregular, natural boundaries, formed by continental coastlines and short connections between the oceans (Spilhaus and Snyder, 1991).

Indirect aspects of world map projections are useful to focus attention on a particular part of the Earth's surface. As the region of interest is placed in the least distorted area of the map, it can be correctly related to its immediate surroundings. Yet parts of the world that are further away from the central region will be severely distorted. This may be experienced as disturbing, especially when it concerns areas of which the shape is well known from conventional maps. Bartholomew developed the Nordic to obtain an optimal representation of Europe, and of routes in the Atlantic, Arctic and Indian Oceans. By putting Europe in the centre of the map, both the Nordic and the Briesemeister projection guarantee a good representation of the Northern Hemisphere. Because the largest part of the continental masses is situated north of the equator (optimisation of the mean linear

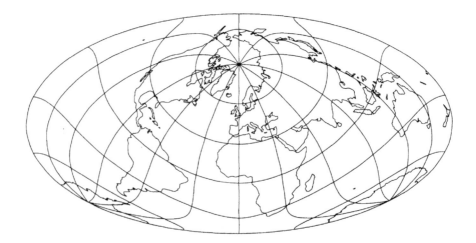

Figure 2.19 Simple oblique aspect of Hammer's polyconic equal-area projection, centred at 45°N, 0°W (Nordic projection).

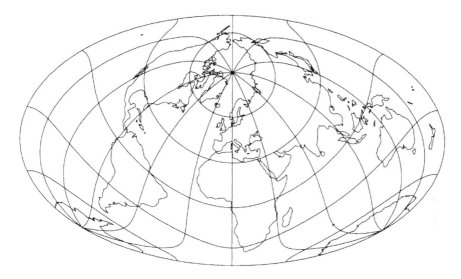

Figure 2.20 Simple oblique aspect of a modified version of the Hammer projection, with a 1.75:1 ratio of the axes instead of the Hammer's 2:1, centred at 45°N, 10°E (Briesemeister projection).

scale distortion over continental area leads to a projection with centre at 25°E, 60°N), mean distortion values are substantially lower than for the normal aspect of Hammer's projection (table 2.3). Yet the continents of the Southern Hemisphere, being pushed closer to the edges of the map, are severely distorted, on the Nordic as well as on the Briesemeister. Both projections are therefore less suited if a good representation of the whole continental area is required.

For a given region of interest, it is theoretically possible to develop an unlimited number of indirect aspects of the same projection, depending on the orientation of the axes of symmetry of the meta-graticule. In general, however, only few of these aspects will represent the world without interruption and/or extreme distortion of the major continents. The calculation of integral distortion values may help to find out which aspects are most interesting. Figure 2.21 shows a first transverse aspect of Mollweide's projection with the North Pole situated at its centre. The map was obtained by minimising the mean linear scale distortion over the continental area. As can be seen, the choice of the greater axis at 25°W, 155°E (rounded off to the nearest 5°) guarantees that, except from Antarctica, no continents are interrupted. Similarly, for a transverse oblique aspect, the longitude of the greater axis and the latitude of the centre of the projection may be determined simultaneously. Doing so for Mollweide's projection yields an aspect with its centre at 45°N, 35°W, a position that is very close to the centre of the well-known Atlantis projection (Bartholomew, 1948) (figure 2.22). The mean linear scale distortion for the Atlantis projection, is almost half of the value obtained for the direct aspect of the Mollweide (table 2.3). Optimisation also yields a second local minimum with similar mean distortion values, centred at 30°N, 80°E (figure 2.23).

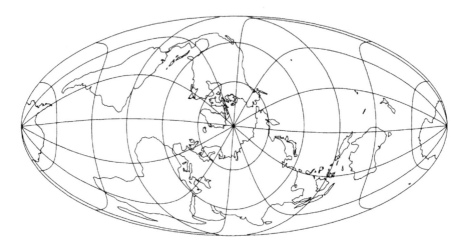

Figure 2.21 First transverse aspect of Mollweide's pseudocylindrical equal-area projection with greater axis at 25°W, 155°E.

In contrast to the Atlantis projection the Indian Ocean is now placed in the centre of the map and the Eastern Hemisphere is represented with less distortion. A disadvantage of this aspect, however, is the interruption of South America.

From the foregoing it is clear that the calculation of integral distortion measures can be useful in choosing a suitable aspect of a map projection. Minimising overall distortion may suggest alternative views, and make the choice of aspect easier and more objective. Of course, indirect aspects of map projections will always introduce visually disturbing distortion in some parts of the map. As has been shown in the previous section, the selection of an optimal projection for a world map requires a careful balancing of a number of geometric qualities that define the characteristic appearance of the graticule. The optimal configuration of meridians and parallels will depend on the way the individual continents are distributed over the surface of the globe. Although a particular map projection may represent the continents well in its normal aspect, there is absolutely no reason to assume that an indirect aspect of the same projection will also yield an appropriate world map. The contrary is true. Because a change of aspect only alters the position of the continents, and has no influence on the distortion characteristics of the projection, the initial configuration of the meta-graticule may eventually become an obstacle to the development of a world map that is meant to focus attention on a particular part of the globe.

For projections that are centred on the equator, the introduction of a pole line usually produces a more balanced representation of the continents (see section 2.5.2). A change of aspect, however, will always place at least one of the major continents close to the pole line of the meta-graticule, thus resulting in an excessive distortion of its characteristic outline. Consequently, projections that represent the pole as a line are seldom used for the development of indirect aspects. It is understood that other qualities of the meta-graticule, not one of them affected

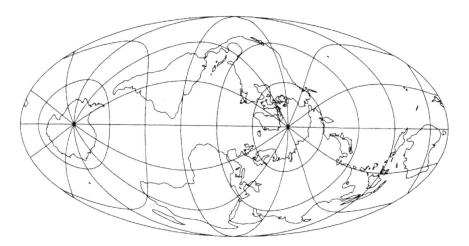

Figure 2.22 Transverse oblique aspect of Mollweide's pseudocylindrical equal-area projection with centre at 45°N, 30°W (Atlantis projection).

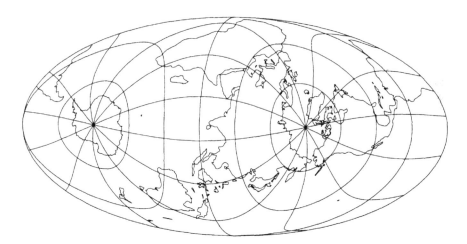

Figure 2.23 Transverse oblique aspect of Mollweide's pseudocylindrical equal-area projection with centre at 30°N, 80°W.

by the change of aspect, may give rise to similar problems. Ideally, for each change of aspect one should alter the characteristics of the meta-graticule, and develop an entirely new projection each time another part of the world has to be placed in the focus of the map. How this can be achieved will be demonstrated in chapter 5.

Of course, no matter how suited for the purpose the use of such a "tailor-made" oblique map projection may be, a considerable amount of distortion near the edges of the map will remain, just like for commonly used map projections. This poses a particular problem that may partly explain why cartographers do not use indirect aspects more frequently. Each individual's mental map of the world is strongly determined by the map projections he or she is most familiar with. Since the majority of map projections has a conventional arrangement of meridians and parallels, with the equator put in the centre of the map, extreme distortion of the polar areas on these projections gradually becomes part of our image of the world, and eventually stops bothering us. Of course, for oblique projections it works the other way. Distorted representations of areas that are located close to the centre on conventional map projections, but that are, through a change of aspect, moved to the edges of the map, will immediately be perceived, and usually disapproved, as they do not correspond with our mental image of these areas. Although the use of indirect aspects offers a unique ability to throw light upon spatial relationships that may be important and cannot be shown on conventional graticules, these psychological barriers are hard to overcome, and stand in the way of a more creative use of map projections.

2.7 PRACTICAL USE AND LIMITATIONS OF LOCAL DISTORTION ANALYSIS

The typical distortion pattern of a projection is related to the appearance of the graticule in its normal aspect, i.e. the meta-graticule. Hence the shape of the meta-graticule is a good criterion for the classification of map projections (see also section 1.4.2). A comparison of the distortion patterns for different classes of world map projections indicates that doubly symmetric projections have more favourable distortion patterns than projections that belong to the conic group. This is, however, not reflected in the mean distortion values obtained for the different graticules (table 2.3). For instance, for Miller's equirectangular cylindrical map projection (figure 3.2), with its strong E–W compression of the equatorial areas, mean values for angular distortion and scale distortion are lower than for most pseudocylindrical projections with equally spaced parallels and pole line. This in spite of the fact that the latter have more balanced distortion patterns, with less variation in local scale factors. These results should not surprise us. Already in 1909, Behrmann compared a number of equal-area projections by calculating the mean angular distortion for each of them (Behrmann, 1909, see also section 2.2). He designated the cylindrical equal-area projection with two standard parallels at ±30° latitude as the best equal-area map for a global representation of the Earth, because it had the lowest mean angular distortion of all projections he considered. A possible explanation for this over-esteem of cylindrical projections may be that integral distortion measures that are obtained by averaging local distortion values do not take account of the spatial variation of distortion values characteristic of each projection. It is therefore wise to use these measures only for the comparison of projections with similar distortion patterns, i.e. for projections that belong to the same class.

    A detailed examination of the doubly symmetric projections that have been introduced in this chapter shows that an optimal representation of the continental area is obtained for graticules with: (i) limited variation of scale along the central meridian, (ii) a 2:1 or slightly smaller ratio of the axes, (iii) a pole line half or less than half the length of the equator, (iv) meridians that do not meet the pole line at very sharp angles. These observations, which are based on quantitative analysis of distortion, and which are confirmed by visual inspection of the graticules that have been analysed, may be useful in deciding about the nature of the graticule when designing new projections for world maps. Of course, they are only valid for a normal (direct) orientation of the graticule. It is not possible to provide guidelines for an optimal configuration of the meta-graticule for indirect aspects, since each change of aspect alters the position of the major continents. All will depend on the location of the area that has to be brought into focus. As indirect aspects usually place one of the more familiar continental shapes close to the edges of the map, the use of pole lines will be avoided in general.

The above observations should be interpreted as general recommendations only, and still leave sufficient flexibility for graticule adjustment. A moderate variation of scale along the central meridian, however, implies that optimal projections for world maps will never be of the conformal or the equal-area type, but will have intermediate distortion characteristics. As has been explained before (see section 2.4.2), it is exactly the balancing of angular and area distortion, and the absence of extreme distortion values, that will guarantee a better representation of the continental area. On the other hand, it is not because an "optimal" representation of the continents is obtained for graticules with a geometry similar to the one described above, that projections of this type will always be the best choice. On the contrary, continental shape is not the only concern, and some geometric properties that are imposed by the purpose of the map may overrule those that favour an optimal distribution of distortion (see chapter 6). Also for some applications a special distortion property (e.g. the equal-area property) may be more valuable than a realistic portrayal of the continents.

Once a decision has been made on the general appearance of the graticule, it is still very difficult to make an optimal choice from a group of candidate projections, solely on the basis of the integral distortion measures that have been described up till now. The mean values for scale distortion prove to be very alike for projections that satisfy similar geometric conditions, even if the graticules of these projections show marked differences in their representation of the continents (table 2.3). Consequently, the question rises if integral distortion measures that are based on local distortion analysis are really appropriate for the evaluation of distortion as it occurs on small-scale maps and, in particular, on maps that represent the whole surface of the Earth.

Indeed, as previously stated, for small-scale maps stress should lay on the reduction of distortion as it is perceived by the map user. This means that the continents should not be interrupted, and that the distortion of large shapes should be kept as low as possible. Tissot's theory of distortion, however, is a local theory, and does not allow the quantification of distortion in the large. More recently, several attempts have been made to quantify map projection distortion for distances, areas and shapes of finite size. In the next chapter, the measurement of finite distortion will be treated in detail. Different methods for the measurement of distortion at the finite scale will be reviewed and evaluated in terms of their ability to characterise overall distortion on small-scale maps. Improved measures of finite distortion will be proposed and applied to a subset of the projections that have been discussed in the present chapter.

CHAPTER THREE

Map projection distortion at the finite scale

Today, map projections which are neither conformal nor equal-area are becoming increasingly popular in small-scale cartography. As was demonstrated in the previous chapter, the favourable distortion patterns of these projections result from balancing the distortion of angles and area, and lead to a better representation of large areal features. This is confirmed by several earlier studies which compare the shape of familiar figures such as the human head (Reeves, 1910), spherical squares (Stewart, 1943; Chamberlin, 1947), spherical triangles (Fisher and Miller, 1944) and spherical circles (Tobler, 1964), after being transformed by different map projections. As world maps influence to a great extent our perception of geographic space, a good representation of continental area and shape should be a major concern. However, in spite of the importance of confronting the map reader with an image that minimises the risk of inducing erroneous impressions about the size and shape of the major landmasses, only few map projections have been developed with the explicit intention of producing an image of the continents that closely resembles their appearance on the globe.

If we assume that the "normal" or "true" shape of a continent against which we make judgements is the one seen if looking at the globe orthogonal at the continent's centre, than the orthographic azimuthal projection, which represents a part of the globe as we would see it from a considerable distance, would occur to be the most obvious choice. Yet the orthographic projection already shows severe distortion at a relatively small distance from its centre. As a consequence, an accurate representation of the whole surface of the Earth is only possible if each major continent is represented with respect to its own centre, yielding a composite world map that consists of different oblique orthographic components (Dent, 1987). Although probably useful for some thematic applications, the discontinuous character of such a poly-centred projection makes it less suited for general-purpose world maps.

3.1 IN SEARCH OF THE RIGHT LOOKING WORLD MAP

In 1963, A.H. Robinson developed a new map projection especially designed for general-purpose world maps, in pursuance of a request by the Rand McNally Company, one of the leading cartographic enterprises in the United States. After considerable analysis of their needs, Rand McNally decided that they needed a map projection with a particular set of attributes of which the most important were (Robinson, 1974):

1. Uninterrupted representation of the entire Earth with each continent appearing as one unit, i.e. located in one section of the projection.
2. Least possible appearance of shearing and approximation of correct relative size for each continent.
3. A graticule which appears simple and straightforward and which is suitable for use by readers of all ages.
4. An overall shape suitable for a sheet format with a proportion of approximately 1.4:1.

A study of existing map projections indicated that all seemed to violate at least one of the listed requirements. Hence the decision was made to develop a new graticule. Based upon Rand McNally's demands, and a good knowledge about map projection distortion as it occurs on well-known projections, Robinson decided in favour of a pseudocylindrical projection with a pole line more than half the length of the equator, and a ratio of the axes not above 2:1. However, instead of proceeding in the traditional way, i.e. by developing map projection formulas starting from a mathematical description of the meridional curves and *a priori* specified distortion properties, he took a rather innovative approach. Starting with an arbitrary projection of the requested type, Robinson gradually adjusted the length and the spacing of the parallels, each time drawing a new graticule, plotting the continents, and judging the result. He repeated this process until it became obvious to him that further adjustment would produce no further improvement in the portrayal of the continents. As Robinson states "the approach is essentially artistic in that the resulting projection is an interpretation distilled from the experience of the author" (Robinson, 1974). The Robinson projection, originally also called the *Orthophanic* (right appearing), is very pleasing to the eye (figure 3.1), yet its construction method is highly subjective, not reproducible, and does not lend itself to a routine production of small-scale maps with a low distortion of large areas.

A somewhat similar approach was taken by Baranyi (1968), who presented seven non-equal-area projections of the pseudocylindrical type, yet with meridians becoming gradually closer away from the central meridian on five of them. The spacing of the parallels varies on all projections, sometimes even increasing and again decreasing from equator to pole. The outer meridians consist of circular arc segments, of which the central arc is centred on the equator, the other arcs on the central meridian or its extension. Other meridians are constructed by dividing the parallels, using the same proportions as for the equator. On some projections the poles are points, on others they are represented as lines half or more than half the length of the equator. Both Robinson's and Baranyi's work date from a time when computers were not yet commonly used in map projection research. Their most important argument against the common approach to map projection development was that by starting from mathematical formulas that describe the general appearance of the graticule, and then adjusting a few parameter values to avoid extreme distortion in some parts of the map, the possibilities to alter the appearance of the continents are relatively limited. That is why they chose to proceed in an empirical, non-mathematical way.

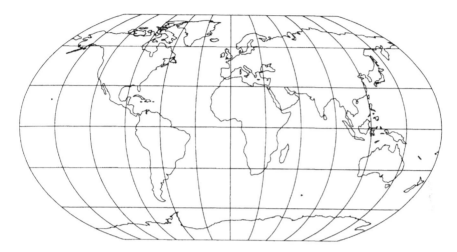

Figure 3.1 Robinson's pseudocylindrical projection with pole line.

The adoption of computers, has created a whole range of new possibilities for the handling of all sorts of problems related to map projection. One of these is the use of power series to approximate map projection coordinate transformations, applying a least-squares fit to a large number of graticule intersections. This may be done to increase the efficiency of data transfer between maps of different projections (Snyder, 1985), or to derive empirical formulas for projections that are obtained by geometric construction, or other non-mathematical techniques. Baranyi and Györffy (1989), for example, published closely fitting empirical formulas for Baranyi's two most popular projections. Canters and Decleir (1989, p. 143) approximated Robinson's projection by two fifth-order polynomials, using a weighted least-squares approach, originally proposed by Wu and Yang (1981), and described in detail by Snyder (1985). It can easily be verified that fifth-order bivariate polynomials are accurate enough to support the construction of large-size world maps (1:5 000 000) in any well-known map projection. Another important area of map projection research in which computers are intensively used is the development of maps with minimum or low distortion, obtained by adapting the parameters of existing map projections to the region to be mapped. While an analytical solution is often quite difficult, if not impossible to achieve, optimised projections can now be developed using iterative solutions. Both least-squares and simplex methods have been used successfully for the optimisation of map projections with a substantial number of projection parameters (see section 5.1).

Given the possibility to approximate an arbitrary projection to a more than reasonable level of accuracy using bivariate polynomials, and to optimise its coefficient values using numerical techniques, there is clearly a potential for a mathematically-based development of new graticules with a low distortion of large areas, without losing the flexibility to accommodate to different appearances of the projection's graticule. In contrast to Robinson's and Baranyi's work, which both can be considered as one-time, non-reproducible efforts, providing a mathematical framework for the development of what might be called *orthophanic* map projections may create opportunities for the production of a new family of low distortion graticules, each with its own properties, in reply to application-specific demands, and optimally fitting the region to be mapped. Accomplishing this is one of the main aims of this study, and is described in detail in the fifth chapter.

3.2   WHY MEASURE DISTORTION AT THE FINITE SCALE?

In order to develop a strategy to minimise the distortion of large areas, one first has to define a measure that is able to quantify the overall distortion of large areas in a proper way. This is not as simple as it may seem. In order for a map to give a portrayal of the continents as realistic as it can be, all continents should appear in true proportion, while the shape of each continent should be as close as possible to its original shape on the globe. Starting from the calculus-based theory of map projections, outlined in the first chapter, a proper quantification and evaluation of distortion in the large is difficult to achieve. While integration of the local areal scale factor (see equation (1.10)) over a region of finite size theoretically produces the region's area on the map (Albinus, 1979), no such relation between local and global aspects of distortion exists for the quality of shape. At infinitesimal scale, shape is preserved if the distortion of scale is equal in all directions, i.e. for conformal projections. As has been shown before, however, conformal projections fail to preserve the shape of large areas as a result of strong variations in scale from one map location to another.

While a careful balancing of angular and area distortion seems to offer the best chances for an overall reduction of distortion, as it is perceived by the map user, the enhanced linear distortion measure (see equation (2.18)), which quantifies the combined effect of angular and area distortion, did not prove to be a good measure for the evaluation of distortion in the large. For example, mean linear scale distortion values for cylindrical projections are not significantly higher than those obtained for doubly symmetric projections (table 2.3). Yet cylindrical projections have less favourable distortion patterns leading to extreme and visually disturbing distortion in different parts of the mapped area. Robinson's pseudocylindrical projection with pole line (figure 3.1) and Miller's equirectangular projection (figure 3.2), for instance, have similar mean linear scale distortion values. A quick look at both graticules indicates that they are clearly not alike in their portrayal of the continents.

A similar remark can be made for pseudocylindrical equal-area projections with pole line. While in the course of the last two centuries numerous projections of this type have been developed, by choosing differently shaped meridians, varying the length of the pole line, and adjusting the ratio of the axes, the mean linear scale distortion value E_{abc}, obtained for these projections, shows very little variation (table 2.3). While this might give the impression that these projections hardly differ from one another in their portrayal of the continents, earlier discussions about the impact of graticule geometry on the distortion pattern of a projection have proven the contrary (see section 2.5). Both examples demonstrate the major

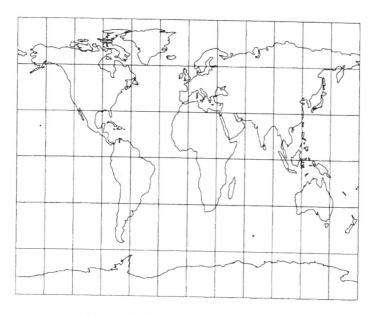

Figure 3.2 Miller's equirectangular projection.

shortcoming of integral measures of scale distortion, that are based on Tissot's classical theory, as a tool for small-scale map projection evaluation. Since these measures do not take account of how distortion varies over the area that is covered by the map, they do not measure the cumulative impact of scale distortion over finite distances. It is needless to say that it is exactly the distortion of finite distances that will determine how well the continents are represented. One may therefore expect to be more successful in evaluating the departure of the continents on the map from their corresponding image on the sphere by measuring map projection distortion at the finite scale.

The measurement of map projection distortion at the finite scale is most easily justified for the evaluation and optimisation of world map projections. Indeed, when the entire world is shown on a map, distortion of the peripheral areas becomes so large that it will be observed by most map users. Reducing this noticeable distortion will obviously be one of the main goals when designing a map projection that is meant for representing the entire world. Nevertheless, the calculation and the minimisation of finite distortion may also be extremely valuable if an optimal reference system has to be defined for an entire continent, or for a region that covers a very large area. This is particularly so if the reference system is to become the geometric base for finite distance calculations, as it is the case in many GIS applications. In the remainder of this chapter, attention will be focused mainly upon global mapping, and upon the use of finite distortion measures for the evaluation and reduction of visible distortion. Chapter 5, however, includes several examples of the use of finite distortion measures for defining optimal projections for continental and regional mapping.

3.3 EARLY STUDIES ON FINITE MAP PROJECTION DISTORTION

One of the first attempts to quantify map distortion in the large was by Fisher and Miller (1944). By defining twenty equilateral spherical triangles on the globe, and comparing these with their mapped versions, obtained by connecting the projected vertices of the original triangles with straight lines, Fisher and Miller were able to calculate the linear scale ratio (map distance over spherical distance) for each side, the areal scale ratio (plane area over spherical area) for each triangle, and the angular difference (plane value minus spherical value) for each pair of sides. From the calculated ratios the maximum, minimum and mean value was derived, as well as the standard deviation. Table 3.1 shows the results obtained for the sinusoidal projection. As can be seen, the mean and the maximum areal scale ratio both differ from one. Knowing that the sinusoidal projection is equal-area this may seem a bit surprising. Yet the explanation for this apparent contradiction is quite simple. The great circles defining the spherical triangles on the globe will generally be mapped as curved lines, which do not coincide with the straight lines that connect the vertices of the triangles on the map. As such both triangle sets are not spatially identical.

As Tobler (1964) points out, however, Fisher and Miller's approach appears to be realistic since the map user's image of a geographic object will always be the one obtained by connecting the set of vertices that define the contours of the object by straight lines. This draws attention to an important issue in the measurement of the distortion of objects of finite size, i.e. the distortion of an object is not only caused by the characteristics of the projection, yet also by the discrete definition of the object boundaries. The more generalised the object boundaries become, the higher the distortion that is caused by the straight line approximation of the great circle arcs connecting the object's vertices will be. When measuring finite distortion, it is obvious that both elements contributing to the distortion should be taken into account. This justifies Fisher and Miller's approach, and explains why a continent-sized region on an equal-area world map will never have its correct area. Similarly azimuths may be incorrectly measured on conformal projections, even though all angles are correctly preserved locally. The latter problem is well known in surveying, and methods have been developed to correct for measurement errors of this kind (Maling, 1992, p. 327).

In the study by Fisher and Miller, the number of points involved was quite

Table 3.1 Finite distortion characteristics for the sinusoidal projection (Fisher and Miller, 1944).

	Linear scale ratio	Areal scale ratio	Angular difference
Mean	1.196	0.767	−12°
Standard deviation	0.521	0.344	39.15°
Maximum	3.991	1.476	108°
Minimum	0.673	0.000	−72°

limited in order to restrict the volume of computational work. With the advent of digital computers, numerical solutions to map projection problems became an interesting and rapidly developing area of research. One of the pioneers and uncontested leader in this field in its earlier days is Waldo Tobler. Taking advantage of the increased computational facilities that became available he conducted a detailed study on finite map projection distortion (Tobler, 1964), and presented least-squares solutions to derive empirical mapping functions for ancient maps (Tobler, 1966b), and to develop new *optimum* projections (Tobler, 1977) (see section 5.1). In his first study on finite map projection distortion, Tobler adopted the approach described by Fisher and Miller (Tobler, 1964). However, instead of working with a small number of *a priori* defined spherical triangles, Tobler evaluated distance, angular and area distortion by randomly generating triplets of latitude and longitude locations. In order to obtain a homogeneous distribution of sample locations the convergence of the meridians was taken into account. This can easily be accomplished by using the following formulas for the random selection of latitude and longitude (Albinus, 1981):

$$\lambda = 2\pi r - \pi \tag{3.1}$$

$$\phi = \arcsin(2r - 1) \tag{3.2}$$

with *r* a random number between 0 and 1, produced by a random generator.

In accordance with Robinson's idea of restricting the computation of distortion to the continental landmasses (Robinson, 1951) (see section 2.3), Tobler also forced the distribution of randomly selected points to conform to the observed frequency of continental area within each five degree increment zone of longitude and latitude. Tobler's justification for calculating distortion values from a large number of randomly generated triangles, instead of using a small set of fixed triangles, was that the map user is not concerned with the distortion of specific distances, angles or areas, but is mainly interested in the overall error. By using a sample size that is large enough to obtain statistically valid results, random sets may give a reliable estimate of the overall distortion.

Applying the method described above, Tobler calculated mean linear scale ratio, mean areal scale ratio, and mean angular difference for a variety of map projections and for areas of different size, using samples of 300 or 500 random triplets, depending on the size of the area (Tobler, 1964). From a quick comparison of his results with mean linear scale ratios, obtained by calculating the mean of all maximum and minimum scale distortion values of Tissot's indicatrix for a set of locations within the area of interest, Tobler concludes that finite distortion values correspond well to what one would obtain on the basis of infinitesimal distortion analysis for areas as large as the United States. Although no detailed comparison was performed, Tobler also states that his results seem to indicate that this correspondence no longer applies for maps of the entire world. This confirms what has already been said about the limitations of infinitesimal measures of distortion if applied to large areas, and may explain the unsatisfactory results that have been obtained with these measures earlier in this study.

3.4   GLOBAL MEASURES OF THE DISTORTION OF FINITE DISTANCE

Next to the numerous efforts that have been done to quantify the distortion of infinitesimal length (see sections 2.1, 2.2), some authors have defined measures to quantify the distortion of finite distances. Again the general idea behind the development of these measures is that all types of distortion that occur on maps can be considered the result of a scale distortion that changes continuously with location and direction. Hence global measures that quantify the distortion of finite distances will provide a good overall idea of the extent to which the plane map differs from the surface that is projected.

Claiming that Airy's measure of distortion involves only local scale factors, and may not give an appropriate criterion for the quantification of finite scale error, Gilbert (1974) quantifies map distortion by a normalised mean-squared error that is defined as follows:

$$E_G = \mathrm{E}\left(\left|s - s'\right|^2\right) \Big/ \sqrt{\mathrm{E}\left(s^2\right)\mathrm{E}\left(s'^2\right)} \tag{3.3}$$

with s the distance between two points on the generating globe, and s' the corresponding distance on the map. The expectation $\mathrm{E}(|s - s'|^2)$ can be obtained by averaging $(s - s')^2$ over a sufficiently large number of randomly selected pairs of points. It is a simple index of the extent to which the mapping departs from isometry. The denominator in (3.3) is a normalising factor that is used to make the error index scale-invariant. In his paper, Gilbert presents a perspective azimuthal projection that minimises the error index, and found it to be close to Lambert's azimuthal equal-area projection for spherical caps up to the size of a complete hemisphere.

Peters (1975) criticises Gilbert's index, and states that the measurement of the distortion of a particular finite distance should satisfy the following conditions:

1.   If no distortion occurs the distortion value should be zero. The stronger the distortion, the higher the distortion value should be.
2.   The distortion value should only depend on the relative difference between the spherical and the projected distance, not on the absolute difference.
3.   The distortion value should be equal for enlargements and reductions by the same factor, in other words, a distance which is doubled in length should get the same distortion value as a distance which is represented at half of its original length.
4.   The distortion value may not become infinite, since in that case the mean distortion of a large number of distances will also be infinite, and no objective comparison of different projections or different optimisations of map projection parameter values will be possible.

Peters defines the distortion e_s of a finite distance s as follows:

$$e_s = \frac{|s - s'|}{|s + s'|} \tag{3.4}$$

where s and s' have the same meaning as above. As can be verified (3.4) satisfies all four conditions. The value of e_s varies between 0 (no distortion) and 1, yet the maximum value (which corresponds to infinite stretching) is never attained.

The value of e_s can be converted to a corresponding scale factor k (= distortion of a distance of unit length) as follows:

$$\frac{|1 - k|}{|1 + k|} = e_s \tag{3.5}$$

which leads to the following quadratic equation:

$$(1 - e_s^2)k^2 - 2(1 + e_s^2)k + (1 - e_s^2) = 0 \tag{3.6}$$

and yields two reciprocal solutions:

$$k_1 = \frac{1 + e_s}{1 - e_s} \tag{3.7}$$

and

$$k_2 = \frac{1 - e_s}{1 + e_s} \tag{3.8}$$

with $0 < k_2 \le 1 \le k_1$. The scale factors k_1 and k_2 respectively correspond with the enlargement and reduction that will occur for a given finite scale distortion e_s.

To characterise the overall distortion of a projection, Peters averages the value of e_s for 30 000 distances, randomly chosen over the entire surface of the Earth:

$$E_P = \frac{1}{30000} \sum_i \frac{|s_i - s'_i|}{|s_i + s'_i|} \tag{3.9}$$

From this average value, corresponding scale factors K_1 and K_2 can be derived as before. A few years later, Peters refined his method by only considering distances connecting points on the continental surface (Peters, 1978). Since pairs of points are randomly selected, spherical distances range from 0 to π. There is, however, one difficulty involved. If the arc of the great circle, along which the shortest distance s between two points on the sphere is measured, passes through the map's bounding meridian, the straight-line connection between both points on the map does not correspond to the shortest distance, yet to the complementary distance $2\pi - s$, measured along the same great circle. This will be the case for all pairs of points for which one observes a longitudinal (or meta-longitudinal) difference greater than π on the map. Peters decided to leave these pairs of points out in his analysis.

Peters calculated the mean finite scale distortion E_P for a number of well-known projections. To verify the statistical validity of his approach he recalculated the value of E_P for different samples of points. Converting E_P to a proportional error ($100K_1 - 100$) he observed a maximum deviation in his results of 4% which he considered acceptable. Table 3.2 shows the results of Peters' analysis, obtained by calculating the mean distortion E_P for 30 000 distances connecting points on the continental surface. His so-called *Entfernungsbezogene Weltkarte* (figure 3.3), which has the lowest value for E_P of all equal-area projections he considered, is obtained through optimisation of the parameters of the Hammer–Wagner projection (see also section 4.1.2). Of all non equal-area projections involved in the analysis only the Winkel–Tripel projection has a value for E_P which is lower than the value obtained for the optimised Hammer–Wagner graticule. Converting E_P to a scale factor, the difference between both map projections seems insignificant. From this Peters concludes that little can be gained by giving up the equal-area property, and pleads for the use of equal-area projections for world maps.

Peters' conclusion is somewhat surprising, given the present tendency to use projections that are neither conformal nor equal-area for the mapping of world-wide phenomena. As has already been mentioned before, the analysis of local distortion patterns indicates that equal-area projections have substantially higher angular distortion (see section 2.4.2), while detailed visual examination of different equal-area and non equal-area graticules clearly shows that this increased angular distortion leads to a less favourable representation of large shapes. Peters also carried out a local analysis of distortion of which the results seem to agree well with earlier studies (table 3.3).

Table 3.2 Mean distortion of distance for a number of world map projections (Peters, 1978).

	E_P	K_1	K_2
EQUAL-AREA PROJECTIONS			
Entfernungsbezogene Weltkarte	0.0521	1.110	0.901
Hammer–Wagner	0.0544	1.115	0.897
Mollweide	0.0566	1.120	0.893
Hammer–Aitoff	0.0575	1.122	0.891
Sinusoidal	0.0641	1.137	0.880
Behrmann's cylindrical equal-area	0.0667	1.143	0.875
OTHER PROJECTIONS			
Winkel–Tripel	0.0517	1.109	0.902
Aitoff	0.0553	1.117	0.895
Kavrayskiy VII	0.0557	1.118	0.894
Cylindrical equidistant with standard latitude 30°	0.0663	1.142	0.876
Mercator	0.1270	1.291	0.775

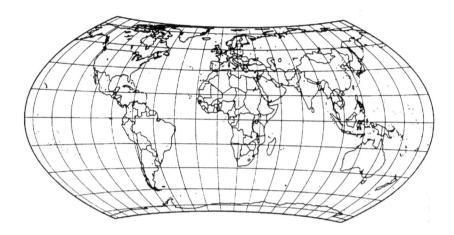

Figure 3.3 Peters' *Entfernungsbezogene Weltkarte* (from Peters, 1978, p. 110; reprinted with permission from Aribert Peters).

Claiming that his equal-area *Entfernungsbezogene Weltkarte* has a mean finite scale distortion value which is almost identical to the value obtained for the Winkel–Tripel projection, he concludes that the equal-area graticule is to be preferred, since the Winkel–Tripel has an area distortion which amounts to 54% at

Table 3.3 Results of Peters' local analysis of distortion (Peters, 1978).

	Mean finite scale error (%) (*)	Mean angular distortion (°)	Areal scale error at latitude 60° (%)
EQUAL-AREA PROJECTIONS			
Entfernungsbezogene Weltkarte	11.0	23.0	0.0
Hammer–Wagner	11.5	25.6	0.0
Mollweide	12.0	28.3	0.0
Hammer–Aitoff	12.2	29.1	0.0
Sinusoidal	13.7	33.5	0.0
Behrmann's cylindrical equal-area	14.3	25.1	0.0
OTHER PROJECTIONS			
Winkel–Tripel	10.9	17.4	54.0
Aitoff	11.7	25.0	22.0
Kavrayskiy VII	11.8	16.6	63.0
Cylindrical equidistant (standard latitude 30°)	14.2	13.2	99.0
Mercator	29.1	0.0	300.0

(*) $100K_1 - 100$

the 60° latitude. What he does not mention is that the mean angular distortion value of his optimised graticule is significantly higher than for the Winkel–Tripel projection. A close inspection of the projection also shows that it has substantially more E–W stretching in the higher latitudes than the Winkel–Tripel (figure 1.9), which is typical for an equal-area graticule with a pole line more than half the length of the equator (see section 2.5.2).

It is clear that one should interpret Peters' statement about equal-area projections with caution, and first find out if the distortion measure which he proposes is really appropriate for measuring distortion in the large. The fact is that Peters' distortion measure does not succeed well in discriminating between map projections that have strongly different distortion characteristics. With the exception of Mercator's projection, all projections in Peters' analysis have a mean finite scale error that varies between 10.9% and 14.3%, while the angular and area distortion for these map projections, their graticule geometry, and therefore their representation of the continents, are very different. As such it is doubtful if Peters' measure of distortion, as it has been proposed originally, provides a good criterion for the optimisation of existing map projections, or for the development of new graticules that succeed in attaining a proper balance between angular and area distortion, resulting in a good representation of the major continents.

3.5   SOME SHORTCOMINGS OF PETERS' MEASURE OF DISTORTION

Albinus (1981) discusses Peters' method of evaluating map projection distortion in detail. One of his remarks is concerned with the way random distances are selected. As previously stated, Peters rejects all pairs of points for which the great circle arc along which the shortest distance on the globe is measured cuts the map's outer meridian. Albinus argues that by doing so more weight is given to areas in the middle part of the map, while distortion is most severe near the map's extremities. According to Albinus, this leads to an understatement of real map error. Instead of excluding pairs of points (p_1, p_2) that are located on opposite sides of the outer meridian, he suggests to measure the map distance between these points across the outer meridian. This can be achieved by calculating the location p_3 where the orthodrome passes the map's outer meridian on the globe, deriving the map coordinates of all three points p'_1, p'_2, p'_3 (in fact the point of intersection corresponds with two map locations, $p'_3{}^+$ and $p'_3{}^-$, one on either side of the map), and adding the distances $p'_1 p'_3{}^+$ and $p'_2 p'_3{}^-$ (assuming that p'_1 is the point with the positive x-coordinate).

Albinus calculated the mean finite scale distortion for the entire surface of the Earth for ten well-known map projections, with and without considering pairs of points for which the distance on the map is measured across the outer meridian (table 3.4). As could be expected, for most projections the mean distortion values are substantially increased if all pairs of points are included. Yet the ranking of the projections is hardly changed and remains a bit awkward considering the general outlook of each of the projections. Although Albinus did not take into account the distribution of the continents, and calculated a mean distortion value for the entire surface of the Earth, the higher distortion value for the cylindrical equidistant projection, as compared to the cylindrical equal-area projection, is not what one would expect. Also the relatively moderate distortion value for the sinusoidal projection, with its excessive compression of area in the higher latitudes and along the outer meridian, raises some questions about the fitness of the mean finite scale distortion, as it is defined here, as a criterion for evaluating overall distortion on small-scale maps.

Assuming that the mean finite scale distortion value is calculated to find out how well a projection succeeds in representing the continents (or the entire surface of the Earth) in a realistic way, it may be questioned if it is wise to calculate this value on the basis of randomly selected distances with a length ranging from 0 to π, as it is done by Peters and Albinus. Indeed, since most map projections have a distortion pattern that includes areas where stretching occurs, as well as areas where map features are compressed, finite distortion is expected to decrease for larger distances due to the increased chance that a compensation of enlargement and reduction will occur along the covered path. Including large distances in the calculation of the mean finite scale distortion will therefore reduce the practical value of the distortion measure. On the other hand, selected distances should be large enough to assure that the systematic elongation or compression of large features in particular parts of the map is well captured.

Figure 3.4 illustrates how the mean finite scale distortion varies with distance for some well-known projections. The graph was constructed by calculating the mean finite distortion value for a random selection of 30 000 distances of 10°, 20°, 30°, ... To obtain a random spatial distribution, distances were selected by

Table 3.4 Mean finite scale distortion for some well-known projections, with and without including spherical distances that intersect the outer meridian (Albinus, 1981).

	Not including distances measured across the outer meridian		Including distances measured across the outer meridian	
	E_P	Mean finite scale error (%)	E_P	Mean finite scale error (%)
EQUAL-AREA PROJECTIONS				
Mollweide	0.0593	12.6	0.0665	14.3
Hammer–Aitoff	0.0660	14.1	0.0762	16.5
Sinusoidal	0.0720	15.5	0.0820	17.9
Lambert's cylindrical equal-area	0.0765	16.6	0.0863	18.9
OTHER PROJECTIONS				
Kavrayskiy VII	0.0518	10.9	0.0540	11.4
Winkel–Tripel (st. latitude 40°)	0.0527	11.1	0.0562	11.9
Francula's polyconic	0.0578	12.3	0.0617	13.2
Aitoff	0.0665	14.3	0.0763	16.5
Cylindrical equidistant (st. latitude 0°)	0.0748	16.2	0.0898	19.7
Mercator's cylindrical conformal	0.1156	26.2	0.1329	30.7

randomly picking a start location as well as an azimuthal angle indicating the distance's direction. Applying spherical trigonometry the end location is then easily found. As can be seen from figure 3.4, the mean finite scale distortion steadily grows until a threshold distance is reached. From then on the distortion value decreases. While the phenomenon is observed for every projection, the threshold distance seems to vary between 30° and 60°, depending on the type of projection. Examination of the distance dependency for other projections that are not included in the graph confirms this. From these findings it is clear that the introduction of a *cut-off* distance, which narrows the distance spectrum used in the calculation of the mean finite scale distortion value, will guarantee that the impact of the spatial variation of local scale factors on the distortion of finite distances is maximally accounted for.

Peters' distortion measure still has another drawback. If the purpose of the measure is to develop, or at least identify, those projections that come closest to representing the Earth's surface as we see it on a globe, then the measure should only account for variations in the shape and the relative proportion of geographic objects. In other words, changing the dimension of the graticule without altering its shape should not affect the measure's value. Peters' measure of distortion does not have the property of scale independence. Consider, for example, the well-known cylindrical equidistant projection with one standard parallel (*Plate Carrée*), which is defined as

$$x = R\lambda$$
$$y = R\phi$$

(3.10)

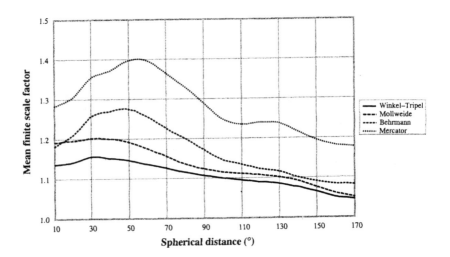

Figure 3.4 Mean finite scale factor as a function of spherical distance for some well-known map projections.

For this projection the local scale factor along the meridians h and along the parallels k is respectively given by

$$h = \frac{dy}{Rd\phi} = 1 \tag{3.11}$$

$$k = \frac{dx}{R\cos\phi d\lambda} = \frac{1}{\cos\phi} \tag{3.12}$$

Thus the nominal scale is preserved along the equator, as well as along each meridian. All other distances are stretched. Yet it is clear that scale variation can be minimised by multiplying the x- and y-coordinate by a factor $k_0 < 1$:

$$x = k_0 R\lambda$$
$$y = k_0 R\phi \tag{3.13}$$

Although equations (3.13) do not define a Plate Carrée *sensu stricto* ($k = k_0 < 1$ along the equator), the new graticule is merely a reduction of the original one, which implies that shape and relative proportion are not changed. However, both graticules have a substantially different mean finite scale distortion according to Peters' measure. This is illustrated in figure 3.5, which represents the mean finite scale factor as a function of the local scale factor k_0 along the equator. Distortion reaches its minimum for $k_0 = 0.9137$. The optimal value of k_0 is, of course, projection dependent. As can be seen, for the Plate Carrée the mean finite scale factor increases in a linear manner for values of $k_0 > 1$. This will, of course, be the case for all projections that have no local scale factors smaller than 1. For these projections a reduction of the mean finite scale factor can only be accomplished by reducing the dimension of the original graticule ($k_0 < 1$). For other projections stretching as well as compressing distances, the optimal value of k_0 may happen to lie above 1. If it is strictly the intention to use the mean finite scale factor as a criterion for evaluating how well the configuration of the continents is preserved on different map projections, their graticules should be scaled until the distortion value reaches its minimum. Only then will it be possible to compare the values obtained for different projections in a scale-independent way.

3.6 FINITE SCALE DISTORTION FOR WELL-KNOWN MAP PROJECTIONS

Taking into account the suggested improvements of Peters' original measure, the overall distortion of finite distances has been evaluated for a small subset of the map projections that have already been subjected to a local analysis of distortion earlier in this study (see section 2.4). Mean finite scale distortion values were calculated on the basis of distances which have both their start and end point located on the continental surface, excluding Antarctica. Since for most projections

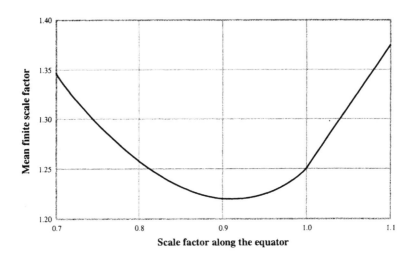

Figure 3.5 Mean finite scale factor as a function of the scale factor along the equator for the Plate Carrée.

(except for cylindrical graticules) distance distortion starts to decrease once great-circle arcs become larger than 30°, this distance was chosen as the cut-off value (see section 3.5). The restriction to a maximum spherical extent of 30° guarantees that distance relationships between areas where the land masses are close (e.g. Asia – Indonesia – Australia) are taken well into account in the measurement of the mean finite distortion value. It should be mentioned, however, that the proposed strategy may slightly underestimate overall distortion for cylindrical projections, as for these projections the distance-related decrease of the mean finite scale distortion value is only observed for distances exceeding 50°.

The algorithm that was used for the random generation of distances produces a set of spherical arcs with a rectangular distance frequency distribution, and works as follows:

1. Select a random distance between 0 and 30°.
2. Independently select a random longitude and latitude for the starting point.
3. If the starting point is situated on the continental area
       then continue
       else go back to 2.
4. Select a random azimuthal angle between 0 and 360°.
5. Calculate the position of the end of the line from the distance, the position of the starting point, and the azimuthal angle.
6. If the end of the line is situated on the continental area
       then accept the line and calculate the finite distortion
       else go back to 2.

This procedure is repeated until the required number of distances is selected and evaluated. Since the computational effort to obtain a mean finite scale distortion value using this approach is quite substantial, tests were performed to find out how many distances are needed to produce a stable result. Although Peters worked with 30 000 distances, a set of only 5000 distances proved more than sufficient to obtain reliable distortion values. In the analysis that is presented here all calculations are based on 5000 distances, randomly selected over the continental surface. For pairs of locations more than 180° in longitude (meta-longitude) apart map distances were calculated across the outer meridian. Graticule dimension was properly adjusted to obtain scale-independent distortion values (see section 3.5).

Table 3.5 shows the mean distortion of distance (E_s), calculated as explained above, the mean finite scale factor (K_1) derived from it, and the adjusting scale factor k_0 for all map projections included in the analysis. Projections have been arranged from the lowest to the highest mean distortion value. As can be seen the Winkel–Tripel projection with standard parallels at ±40° latitude has the lowest mean finite scale distortion of all projections treated in the analysis. The Aitoff–Wagner, which closely resembles the Winkel–Tripel (see section 4.1.2), is the second best, followed by Wagner VIII, Kavrayskiy VII, and Robinson. All five

Table 3.5 Mean distortion of distance (E_s), mean finite scale factor (K_1), and scale adjustment factor (k_0) for the studied map projections.

	E_s	K_1	k_0
Winkel–Tripel (st. latitude 40°)	0.0663	1.142	0.994
Aitoff–Wagner	0.0693	1.149	0.995
Wagner VIII	0.0694	1.149	0.961
Kavrayskiy VII	0.0707	1.152	1.001
Robinson	0.0716	1.154	1.041
Hammer–Wagner	0.0757	1.164	0.964
Wagner II	0.0761	1.165	0.952
Eckert IV	0.0769	1.167	0.960
Wagner VI	0.0813	1.177	0.945
Wagner I	0.0822	1.179	0.956
Wagner IV	0.0823	1.179	0.994
Eckert VI	0.0828	1.181	0.955
Cylindrical equidistant (st. latitude 30°)	0.0831	1.181	0.984
Mollweide	0.0905	1.199	0.960
Aitoff	0.0912	1.201	0.950
Cylindrical equal-area (st. latitude 30°)	0.0922	1.203	0.971
Hammer–Aitoff	0.0966	1.214	0.983
Cylindrical equidistant (st. latitude 0°)	0.0979	1.217	0.912
Miller II	0.1023	1.228	0.856
Miller I	0.1093	1.246	0.833
Sinusoidal	0.1161	1.263	0.975
Cylindrical equal-area (st. latitude 0°)	0.1199	1.272	0.973
Mercator's cylindrical conformal	0.1349	1.312	0.788

projections have intermediate distortion characteristics and show the least distortion in the middle latitudes, where the concentration of continental area is the highest. Local analysis of distortion, as performed in chapter 2, also produces very low distortion values for these graticules (see table 2.3). Nevertheless, some clear differences can be observed between the results obtained with local and finite distortion measures. While the results of local distortion analysis indicate that equal-area graticules have considerably higher scale distortion than graticules that are neither equal-area, nor conformal, finite analysis of distortion strongly reduces this contrast between equal-area and non equal-area projections. Although it is still true that equal-area projections have higher distortion values than intermediate projections with similar geometric features, another, clearly more important characteristic of graticule geometry seems to interfere.

Representing the pole as a line instead of a point proves to have a much greater impact on finite distortion than the decrease of parallel spacing, which is usually imposed to make the graticule equal-area, or at least to reduce the amount of area distortion. The first twelve projections in table 3.5 all have a pole line which is half or close to half the length of the equator. The first five are non equal-area projections, among the other seven are five equal-area projections. Cylindrical projections and projections that represent the pole as a point both have higher distortion values, irrespective of the fact if they are equal-area or not. This strong impact of the way the pole is represented corresponds well with how map projection distortion is actually experienced. In present-day cartographic practice, map projections that represent the pole as a point are hardly used for world maps because of the excessive shearing in the high latitudes. Also the use of cylindrical projections is decreasing, especially since professional cartographic associations have started protesting against it.

While overall distortion values that are produced by local distortion analysis are the lowest for non equal-area graticules, Peters' analysis does not contribute much of the extreme distortion on world graticules to the maintenance of the equal-area property. The results obtained with the improved finite distortion measure presented in this study are clearly more subtle for that matter. While showing indeed that the equal-area property has a negative impact on overall distortion, they also suggest that other geometric qualities of the graticule are equally important in the shaping of the Earth's surface, and may strengthen, weaken or even override the consequences of giving up or imposing the equal-area property. The importance of graticule geometry has already been discussed in detail in the previous chapter (see section 2.5), yet its impact on overall continental distortion is not well captured by the mean local scale distortion E_{abc}, and the mean finite scale distortion E_P as defined by Peters. The presently proposed measure seems more sensitive to changes in the geometry of the graticule, which proves that it takes better account of the spatial characteristics of the projection's pattern of distortion. This sensitivity also leads to an increased ability to discriminate between different projections on the basis of their overall distortion.

Although the results obtained with the newly proposed measure are promising, and suggest that the measure may be appropriate for the optimisation of standard map projections, and for the development of new graticules with low distortion of large areal features, it would be interesting to investigate first how the measurement of finite scale distortion, as it has been defined in this study, relates to both qualities we expect from a small-scale map, i.e. the maintenance of true continental proportions, and the proper representation of continental shape. Since equal-area graticules (which maintain true continental proportions) are known to introduce excessive shear, area distortion and shape distortion at the finite scale will probably be related to one another in a similar way as area distortion and angular distortion at the infinitesimal scale. One may also expect the distortion of finite distances to be minimal for intermediate projections that show no extreme deviation from true area proportions and no extreme distortion of shape. To be able to verify this, the relative distortion of area, as well as the distortion of large shapes, should be measurable for each projection.

3.7 MEASURING THE RELATIVE DISTORTION OF AREA

The relative distortion of area is fairly easy to quantify. In his pioneering study on finite map projection distortion, Tobler (1964) already proposed to quantify area distortion by averaging the areal scale ratio (plane area over spherical area) for a large set of randomly generated triangles of different size covering the area of interest (see section 3.3). Alternatively the area distortion for each triangle may be calculated using the equivalent of Peters' distortion measure E_P (equation (3.4)), where plane area S'_i and spherical area S_i are substituted for plane distance s'_i and spherical distance s_i, and then averaged for the entire set:

$$E_A = \frac{1}{m} \sum_{i=1}^{m} \frac{|S_i - S'_i|}{|S_i + S'_i|} \tag{3.14}$$

This way the distortion measure obtains a number of interesting properties similar to Peters' measure (see section 3.4). Again the mean distortion value should be independent of scale since we are only interested in measuring proportional error, not absolute area distortion. As was already explained, this can be accomplished by adjusting the graticule's dimension until distortion is minimised (see section 3.5). In theory the mean distortion value should be zero for equal-area projections, as well as for up-scaled or down-scaled versions of equal-area graticules. However, if the mean distortion value is calculated for a random set of triangles, equal-area projections will always have a proportional distortion due to the fact that the straight sides of the plane triangles will generally not coincide with the projected sides of the spherical triangles. The discrepancies between the straight-line connection of a triangle's mapped vertices and the actual projection of its bounding arcs will grow as the triangle becomes larger.

Of course, one is not restricted to the use of triangles to calculate the relative distortion of area. Any type of closed polygon can be used, although it is true that calculations will become more tedious as the number of vertices increases. By calculating distortion on the basis of objects with many vertices, however, the polygons that are obtained by connecting the mapped vertices, will less deviate from the projected versions of the spherical objects. Hence the distortion measure will be less influenced by this error of approximation, and will give a true account of the error that is due to variations in local scale caused by the projection.

3.8   MEASURING THE DISTORTION OF SHAPE

Measuring the distortion of shape is less obvious than it may seem. Before attempting to quantify the distortion of shape one first has to agree about what is actually meant by the shape of an object, and how it can be measured or described in an objective way. References to shape distortion in map projection literature are confined to purely qualitative, and somewhat arbitrary statements about the extent to which particular projections distort the shape of the continental surface. Apart from a few studies that have already been mentioned, in which map projection distortion is examined by visually comparing the shape of familiar figures as they occur after having been transformed by various map projections (Reeves, 1910; Stewart, 1943; Chamberlin, 1947), no further attempts to systematically investigate or quantify the distortion of shape as caused by map projection are known to this author. Scientific literature, however, includes many references, mainly by geographers, that explicitly deal with subjects related to shape measurement and shape comparison. Before attempting to measure the distortion of shape caused by map projection, we will therefore examine how shape is defined, and how one has tried to quantify it.

3.8.1   The measurement of shape

According to Blair and Biss (1967) "shape is that quality of an object or form which depends on constant relations of position and distance from all the points composing its outline or its external surface" and further "... a set of properties – compactness, elongation, and others – which are spatially related and combine together". From this definition it is clear that the main problem in defining a measure of shape lies in making sure that the measure does not include more or less than shape. Indeed, most indices that are found in literature only measure a sub-property of shape. On the other hand, many indices are not invariant to size and/or orientation and, as such, measure more than shape itself.

Theoretical considerations about the concept of shape and its use in geography are outlined by Bunge in his *Theoretical Geography* (Bunge, 1966). Bunge demonstrates that it is impossible for any single measure to define shape unambiguously. He introduces a method of measuring shape that is based on the theorem that any simply connected shape can be matched by an equilateral polygon of any number of sides. Provided that a given object is approximated to a desired level of accuracy by such a polygon, Bunge illustrates that its shape can be described by a unique set of indices, which are all obtained after summation of distances or squared distances between selected vertices of the polygon. This agrees well with the intuitive notion that the shape of an object depends on the distances between points that characterise its outline, and that are psychologically perceived as defining the quality of shape.

Bunge's indices are unaffected by size, orientation and location. They define shape without any ambiguity, i.e. it is possible to go backwards and forwards, from shape to indices and from indices to shape. The biggest problem with Bunge's method however is that an accurate measurement of shape requires a large number of parameters (the number of vertices of the approximating polygon minus two). Moreover, the use of the method is restricted to simply connected shapes. As such Bunge's method has little practical meaning outside shape classification.

At first the impossibility to measure all features of shape by way of a single index seems rather discouraging. Fortunately, we do not require an exact description of shape for the purpose of this study. We are only interested in the differences of shape that are inherent to the mapping process. Of all the properties of shape that have been described in literature, and that have been quantified by single measures, especially the property of compactness deserves our attention. Compactness may be considered as the extent to which an area is grouped or packed around its central point (Blair and Biss, 1967). As such the property is strongly related to our visual impression of geometric forms.

The most compact of all shapes is the circle. Most indices of compactness therefore compare an irregular shape to a circular standard. Miller, for example, expresses his coefficient of compactness as the area of the irregular shape divided by the area of a circle having the same perimeter (Miller, 1953). This measure does not take account of the relative position of points situated along the outline of the object. Hence it is not related to shape as we perceive it. On the contrary, the dependence on the perimeter tends to overestimate small indentations in the outline of the object that are of less importance to overall shape. Haggett already comes much closer to the intuitive notion of compactness. He compares the area of an object with the area of a circle that has the longest axis of the object as its diameter (Haggett, 1965). The only problem with Haggett's measure is that its value strongly depends on the relative position of the two extreme points of the object. It can only give a very crude approximation of the actual degree of compactness.

Boyce and Clark developed another measure which does not contain this weakness (Boyce and Clark, 1964). They measure distance from the centre of gravity of an object to the outside edges along equally spaced radials. They then compute the percentage of each radial distance with respect to the sum of all radials, subtract the percentage each radial would have been if based on a circle, and take the absolute value of the differences. The results of this operation are summed to determine an index which becomes zero for the circle, and increases with non-compactness. The method has the advantage of measuring the compactness of an object to any desired level of accuracy, depending on the number of radials (or the number of points along the outline of the object) involved in the computation. It is also invariant to size and orientation of the object. Figure 3.6 illustrates how the index of Boyce and Clark varies for ellipses with gradually increasing eccentricity.

3.8.2 Defining a proper measure of shape distortion

Since the index of Boyce and Clark describes an object's departure from the circular shape, a similar procedure can be followed to quantify the shape distortion of a spherical circle after it has been mapped onto a plane. Indeed, for a spherical circle with given centre and radius it suffices to: (a) define a number of equally spaced points along its outline; (b) calculate the position of the centre and the peripheral points on the map; (c) measure the map distance between the projected centre and each of the peripheral points; (d) compute shape distortion in the same way as

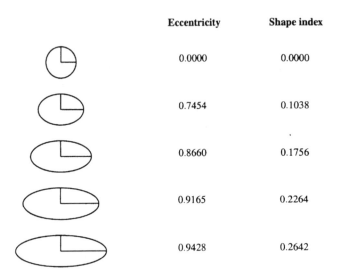

	Eccentricity	Shape index
	0.0000	0.0000
	0.7454	0.1038
	0.8660	0.1756
	0.9165	0.2264
	0.9428	0.2642

Figure 3.6 Boyce and Clark's index of compactness for ellipses with different eccentricity.

Boyce and Clark derive their index. Although not identical to the index of non-compactness (radial distances are measured from the projected centre, not from the centre of gravity; radials are not equally spaced as a result of angular distortion), the distortion measure quantifies the departure from the circular shape. It becomes zero if the circular shape is retained, and increases as shape distortion becomes worse.

Having defined an index of shape distortion for an arbitrary circle, the overall distortion of shape can be measured by averaging the index for a large number of spherical circles, randomly distributed over the land masses. In order to obtain random sets that do not cover too large parts of non-continental area, the maximum size of the circles should be limited. In the following analysis the circular radius has been bounded to 30°. This is also the maximum distance that was used in the analysis of mean finite scale distortion described in section 3.6. Further details about the definition of the circles and the calculation of the shape index are outlined below.

3.9 DISTORTION OF AREA AND SHAPE FOR WELL-KNOWN MAP PROJECTIONS

To obtain an overall measure of the relative distortion of area and the distortion of shape for different map projections, the finite distortion of area E_A (equation (3.14)) and the index of shape distortion, defined in the previous section, have been calculated and averaged for 1000 spherical circles of varying size (circular radius ≤ 30°), randomly generated over the continental area. To ensure that selected circles are mainly located on land the following procedure has been applied:

1. Select the position of the circle's centre (random longitude and latitude).
2. If the centre is situated on the continental area
   then continue
   else go back to 1.
3. Randomly select a circular radius between 0 and 30°.
4. Calculate eight positions along the circle's perimeter for equal increments of azimuth (0°, 45°, 90°, 135°, …).
5. If at least six out of eight positions are situated on continental area
   then continue
   else go back to 3.
6. Calculate eight intermediate positions for azimuthal angles of 22.5°, 67.5°, …
7. Calculate map coordinates for the circle's centre and the 16 points along its perimeter.
8. Determine area distortion and distortion of shape for the projected circle.

The selection of 16 vertices along the circle's perimeter guarantees that the plane polygon that is defined by the position of the vertices in the map can be considered a good approximation of the projected circle. Hence area distortion will be close to zero for equal-area projections, as one should expect, while the measurement of shape distortion will take maximum account of direction-dependent variations in scale. Steps 3 to 5 are repeated maximally 30 times for the same position of the centre. If after 30 iterations no circle has been found for which six out of eight positions on its circumference are located on the land surface then the position of the centre is rejected, and the procedure is restarted with a new location for the centre.

Table 3.6 lists the mean finite scale factor, the relative distortion of area and the distortion of shape for all map projections found in table 3.5. The relative distortion of area is expressed by way of a mean areal scale factor that is derived from E_A in the same way as the mean finite scale factor is derived from E_P (see section 3.4). This makes it easier to get an idea of the magnitude of relative area distortion. As before the graticule's dimension was properly adjusted to obtain a scale-independent measure of the areal error (see sections 3.5, 3.7). Plotting the distortion of shape versus the relative distortion of area (figure 3.7) shows an inverse relationship similar to the relation between angular and area distortion at the local scale (see section 2.4.2, figure 2.4). Shape distortion is the highest for equal-area projections, as could be expected. Relative distortion of area is most prominent for conformal map projections (Mercator), and for projections that have a small distortion of shape (Miller I, Miller II). The Mercator projection still has a considerable distortion of shape, which of course is caused by the strong variation in scale from the equator towards the poles. Projections with intermediate distortion characteristics at the local scale, which have a low value for the mean finite scale factor, also have moderate distortion of area and shape at the finite scale.

Observed differences in the distortion of area and shape for these projections correspond to small shifts in the balance between both types of distortion.

Figure 3.8 shows the relationship between the mean finite scale distortion and the distortion of shape for all equal-area projections included in table 3.6.

Table 3.6 Mean finite scale factor (K_1), mean areal scale factor, and distortion of shape for the studied map projections.

	K_1	Mean areal scale factor	Distortion of shape
Winkel–Tripel (st. latitude 40°)	1.142	1.159	0.098
Aitoff–Wagner	1.149	1.188	0.098
Wagner VIII	1.149	1.188	0.098
Kavrayskiy VII	1.152	1.209	0.093
Robinson	1.154	1.150	0.101
Hammer–Wagner	1.164	1.000	0.139
Wagner II	1.165	1.072	0.127
Eckert IV	1.167	1.000	0.133
Wagner VI	1.177	1.209	0.099
Wagner I	1.179	1.000	0.149
Wagner IV	1.179	1.000	0.142
Eckert VI	1.181	1.000	0.151
Cylindrical equidistant (st. latitude 30°)	1.181	1.323	0.073
Mollweide	1.199	1.000	0.151
Aitoff	1.201	1.098	0.137
Cylindrical equal-area (st. latitude 30°)	1.203	1.000	0.120
Hammer–Aitoff	1.214	1.000	0.155
Cylindrical equidistant (st. latitude 0°)	1.217	1.323	0.088
Miller II	1.228	1.494	0.061
Miller I	1.246	1.606	0.050
Sinusoidal	1.263	1.000	0.176
Cylindrical equal-area (st. latitude 0°)	1.272	1.000	0.143
Mercator's cylindrical conformal	1.312	2.015	0.038

While we can see that a higher distortion of finite distances generally implies a higher distortion of shape, we may also note that cylindrical projections have very favourable shape indices in proportion to their finite scale distortion. Behrmann's cylindrical equal-area projection with standard latitude at 30° has the lowest shape distortion of all equal-area projections in the set, while it is only ranked as the seventh best equal-area projection in terms of its mean finite scale distortion. The cylindrical equal-area projection with correct scale along the equator, which is found at the bottom of table 3.5, and which is known to excessively attenuate the polar areas, has a value for the shape index that is comparable with the values obtained for pseudocylindrical equal-area projections with a pole line. These observations indicate that there is no simple relationship between the distortion of distance and the distortion of shape as it exists at the local scale, at least not for shape distortion the way it has been defined in this study, i.e. as a loss of compactness. It seems that the compactness of an area on the globe is best retained on a cylindrical projection. Low values for the shape index are also observed for cylindrical projections that are not equal-area, although in this case the favourable distortion of shape is coupled to a high relative distortion of area (table 3.6).

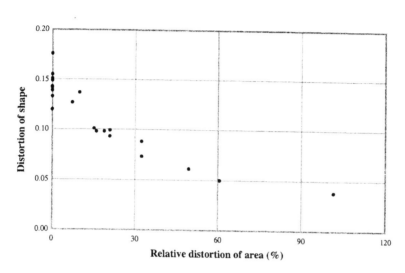

Figure 3.7 Relative distortion of area and distortion of shape for a selection of world map projections (see also table 3.6).

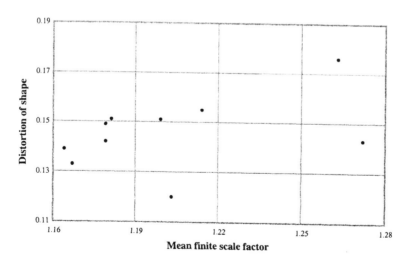

Figure 3.8 Mean finite scale factor and distortion of shape for different equal-area projections (see also table 3.6).

3.10   MEASURING THE RELATIVE CONTRIBUTION OF DIFFERENT KINDS OF DISTORTION

Acknowledging that map projection distortion may present itself in various ways, some authors have proposed distortion measures that assess the combined effect of two, three or even more different kinds of distortion. The classical example of a combined measure of distortion is Airy's first distortion measure (equation (2.4)), which is obtained by averaging two terms that quantify the local distortion of area and shape respectively. As discussed in section 2.1, Airy's measure was generalised later on by introducing different weight factors for both terms in the equation, so that any desired balance between area and shape distortion can be obtained (equation (2.7)). Of course, the principle of weighted averaging can be applied to any combination of distortion measures. In a paper on map projection selection, Bugayevskiy (1982) suggests to compute for each candidate map projection the weighted average of different integral measures of distortion that can be derived from local distortion theory:

$$E_{GEN} = \sum_{i=1}^{n} P_i E_i \bigg/ \sum_{i=1}^{n} P_i \qquad\qquad (3.15)$$

with E_i an integral measure of distortion, and P_i the corresponding weight factor. The measures of distortion he proposes include measures of overall angular, area and scale distortion, which are comparable to the measures defined in chapter 2 (see section 2.2), as well as other measures that describe particular aspects of local distortion, e.g. the mean difference between the curvature of a meridian (or a parallel) on the projection and on the globe, the average departure of a rhumb line and a great circle from a straight line, etc. The significance of each of these measures in the weighting depends on the purpose of the mapping. Although Bugayevskiy indicates that some of the proposed measures may be irrelevant to some applications ($P_i = 0$), he does not offer an objective means to determine the relative weight of each measure. This would present a major difficulty in the practical use of equation (3.15).

One of the main problems in dealing with different kinds of distortion is that all components of distortion are measured in different units and, therefore, cannot be compared directly. Although one can quantify the overall distortion of area and shape for different map projections, and subsequently rank them, based on the values obtained for one of the distortion measures used, it is not possible to say for one particular map projection if it has more area distortion than shape distortion. This incompatibility of units makes it very difficult to properly weigh the different kinds of distortion in a combined measure of the weighted average type, as defined by equation (3.15). To overcome this problem, Laskowski (1998) proposes to calibrate the various distortion values, obtained for each map projection, against an external standard called the *reference projection*. In the case of world maps, he suggests the cylindrical equidistant projection with one standard parallel (*Plate Carrée*) as the reference projection, although he also describes other possible candidates. To motivate the use of the *Plate Carrée*, Laskowski points out its simplicity, and the fact that "... it is equipped (more or less evenly) with all the relevant kinds of distortion ..." (Laskowski, 1998, p. 11). Calibration of distortion values is done by simply dividing the distortion values obtained for each map projection by the corresponding value obtained for the reference projection. Hence, the standardised distortion values for the reference projection itself will all equal 1.0. While the approach can be applied to any possible measure of distortion, Laskowski suggests to compare different map projections by means of three distortion measures, two that quantify the average distortion of area and shape, measured at the infinitesimal scale, and one that quantifies the average distortion of finite distance. Together the standardised values obtained for these three kinds of distortion define the so-called *distortion spectrum* of the projection, which can be represented in a graph in order to facilitate the comparison of different map projections (figure 3.9). To determine the distortion spectrum of a map projection, Laskowski uses the following local measures of area and shape distortion:

$$e_{ar} = (ab - 1)^2 \tag{3.15}$$

$$e_{sh} = \left(\frac{a}{b} - 1\right)^2 \tag{3.16}$$

which both correspond with one of the terms in Airy's first equation. Average values of these two measures are calculated for a set of 5000 randomly selected point locations that cover the globe uniformly. For the average distortion of finite distance he uses Gilbert's distortion measure (equation (3.3)), and calculates it for 5000 randomly selected distances, again uniformly covering the entire surface of the Earth. Laskowksi also defines a *total* or a *cumulative distortion score* for each map projection as the sum of the three components in the distortion spectrum.

The calibration of distortion values against a reference projection has the advantage that the range of possible distortion values becomes more alike for the different kinds of distortion. It should be clear, however, that the distortion spectrum does not allow us to state that the shape distortion on a particular projection is twice as high as the area distortion or the distortion of distance. Although Laskowski claims that the distortion spectrum can be used to make statements of this kind, it is easy to see that the relative weight of the three distortion components in the spectrum will be influenced by the choice of the reference projection. One should thus be careful when interpreting the distortion spectrum of a projection, and be aware that it is based on a relative judgement of distortion only. This does not imply that the distortion spectrum cannot be useful for comparing distortion on various map projections. By choosing a reference projection that occupies an intermediate position between the conformal and the equal-area projections, and calibrating all distortion values against it, it becomes much easier to position a map projection within the distortion continuum than by comparing non-calibrated distortion values for different projections. The distortion spectrum also has a strong visual appeal, which makes it a very suitable device for presenting map projection distortion properties to a non-specialised audience.

Of course, many other distortion measures may be chosen to define the three components in the distortion spectrum. In his study, Laskowski lists numerous

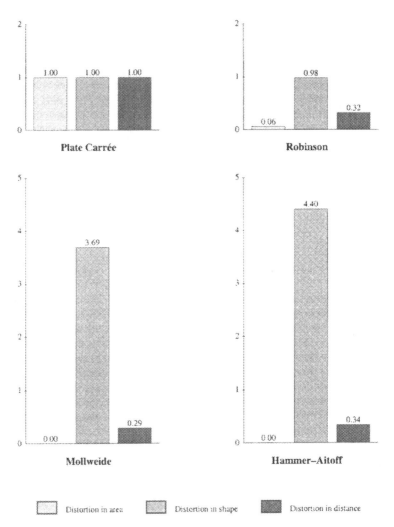

Figure 3.9 Mixed local-global distortion spectra for different map projections, with average distortion values calculated for the whole sphere, and calibrated against the *Plate Carrée* (Laskowski, 1998).

examples of distortion measures, well-known measures as well as previously unpublished ones, which may be used as an alternative to quantify the distortion of area, shape, and distance. Also, other kinds of distortion might be considered to define the spectrum, depending on the types of distortion that are considered most important in the context of the mapping. Indeed, one might as well define the distortion spectrum by combining area, shape and scale distortion at the

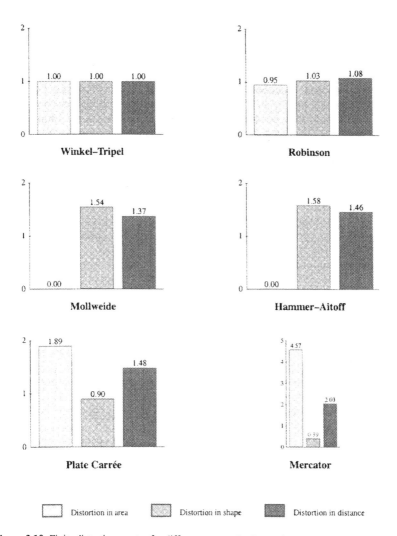

Figure 3.10 Finite distortion spectra for different map projections, with average distortion values calculated for the continental area only, and calibrated against the Winkel–Tripel projection.

infinitesimal level only, or alternatively, by combining area, shape and distance distortion at the finite level. Considering what has been discussed previously in this chapter, the latter option seems especially relevant for comparing the relative contribution of different distortion components on world maps.

As previously stated, the reference projection that is used for calibration will affect the relative height of the different components in the distortion spectrum. Although Laskowski uses the *Plate Carrée* as a reference projection, one could argue if it would not be better to use one of the polyconic or pseudocylindrical projections with low overall scale distortion that were discussed in chapter 2, and earlier in the present chapter (for example, Winkel II, Kavraisky VII, or Winkel–Tripel). Each of these projections is known for its balanced distortion characteristics, and therefore occupies a more or less central position along the distortion continuum. Compared to these "balanced" projections, the *Plate Carrée* has a relatively high distortion of area and, because it is a cylindrical projection, a relatively low distortion of shape (see sections 2.4.2 and 3.9). Using it as a reference projection will reduce the relative contribution of area distortion and increase the relative distortion of shape for all other projections.

Figure 3.10 shows finite distortion spectra for some of the projections that have been analysed in the present chapter. Area, shape and distance distortion values were calculated for the entire continental area (Antarctica not included) by applying the finite distortion measures that were previously defined (see sections 3.6, 3.7, 3.8). Distortion values for each map projection were calibrated against the corresponding distortion values obtained for the Winkel–Tripel projection with standard parallels at ±40° latitude. It is clear from figures 3.9 and 3.10 that the distortion spectrum of a map projection can look quite different if other distortion measures are used, and/or if another reference projection is chosen. One should keep this in mind when using the distortion spectrum for comparative purposes. For the same reason, one should be very careful with the interpretation of total distortion scores, which are obtained by simply adding the three components of the distortion spectrum, or by calculating a weighted average, especially if the total score is used as a criterion for the selection of an overall best projection, or for the development of low-error map projections (see chapter 5). Although the distortion spectrum is a simple and visually appealing tool for summarising the distortion characteristics of a projection, the fundamental problem of weighing different aspects of distortion properly, depending on map function and map use, is far from solved and remains an interesting area for future research.

CHAPTER FOUR

Error-reducing modifications and transformations of map projections

Once a map projection has been selected, its graticule is often modified or transformed to reduce distortion in some part of the map or for the entire map area. In this chapter we will have a look at different methods of transformation that can be applied. A major distinction should be made between continuous and non-continuous map projection transformations. A transformation is said to be continuous if the resulting map represents the spherical surface of the Earth without interruption, and the entire graticule can be constructed from one pair of transformation formulas with properties as defined in section 1.1. Non-continuous transformations use different sets of equations for different parts of the mapped area. By defining a unique transformation for each part of the map, distortion can be further reduced, yet with loss of continuity. Apart from this distinction, the methods of transformation that can be applied are identical for continuous and non-continuous mapping, although in the latter case additional measures may have to be taken to reduce or to avoid the presence of interruptions where different map segments meet. At the end of this chapter, we will also consider some examples of map projection transformations that deliberately introduce distortion, instead of reducing it. Developing maps with substantial scale variation may be useful to enlarge portions of the map that contain detailed information, or that should be brought into focus. Another interesting application of deliberately introducing distortion is the develoment of maps in which the distance between different locations, or the area of the mapping units, is made proportional to some thematic variable.

The simplest transformation of a projection's graticule is accomplished by changing the aspect of the projection. As discussed in section 2.6, a change of aspect does not alter the distortion pattern of the projection, but merely centres it on another part of the Earth's surface, with the intention of reducing distortion in the region to be mapped. There are, however, many other ways of transforming map projections, which do have an impact on the distribution of local scale factors. A common way of transforming map projections is by introducing additional parameters in their transformation formulas. Doing so a more general system of equations is obtained, which includes the original projection as a special case. A transformation of the graticule of this kind is often called a *modification*, although there is no agreement about the exact definition of the term. According to Maling (1968) the use of the word *modified* should be confined to projections in which a single line of zero distortion is replaced by two such lines, a technique which is often applied to reduce distortion towards the edges of the map on conic type projections (see section 1.4.1). Other authors, however, have used the term in a more general sense, to describe all transformations of a projection that maintain the conformal or the equal-area property of the original graticule. Snyder (1987a), for example, uses the term *modified* to describe complex-algebra conformal transformations of the azimuthal stereographic projection (see section 4.2.3). While the projections that are developed in this way are indeed conformal, the typical geometry of the stereographic projection (including the preservation of the circular shape) is lost. Reference to a modified stereographic projection in this case is somewhat misleading, since it may suggest the presence of some properties the projection no longer has.

Alternatively, one might use the term *modification* to refer to all instances of a generalised projection system that are obtained by changing the value of one or more of the parameters in its transformation formulas. Changing parameter values will cause a redistribution of local scale factor values throughout the whole map area, yet without destroying any of the properties attributed to the general projection system. A well-considered choice of parameter values, however, may define a graticule with properties which the general projection does not have. It should be clear that, if applying this definition, the term *modification* no longer refers to the development of new projections, yet only to the adjustment of the graticule of already existing projections through the change of particular parameter values (specialisation). The term *transformation* might then be used to refer to the opposite process, i.e. the creation of a new, more general projection system from an existing projection (generalisation), possibly with loss of some of the properties of the latter.

Applying the above definitions it follows that both the terms *modification* and *transformation* may be used in connection with one general projection system and some of its special cases. The distinction between both terms then only becomes a matter of looking at a particular map projection in two different ways. Miller's general equations for cylindrical projections with adjustable spacing of the parallels, for example, are obtained by *transformation* of Mercator's cylindrical conformal projection. Vice versa, the Mercator projection, as well as the two cylindrical projections proposed by Miller himself, can be obtained by *modification* of the parameter values in Miller's general equations (see section 2.5.1). To simplify things and to avoid any ambiguity, in the following sections we will speak of *modification* in all cases where the graticule of the projection is obtained by changing the parameter values in a set of general map projection equations, which express the map projection coordinates x,y directly in terms of geographical coordinates λ,ϕ on the globe. The term *transformation* will be reserved for a special kind of transformation, which involves the re-projection of map projection coordinates in the plane.

4.1   MODIFICATION OF MAP PROJECTION PARAMETERS

We will now discuss three different types of map projections with adjustable equations: (a) conic type projections, which allow modification of the graticule through changes in the position of the line(s) of zero distortion only, (b) a set of generalised pseudocylindrical and polyconic projections that allow independent adjustment of various graticule characteristics, and (c) a family of generalised map projections that are based on polynomial equations, and that allow maximum flexibility for graticule adjustment. The first two types of map projections are obtained by generalisation of existing map projection equations, and are frequently used in contemporary cartography. The latter are defined independently of any already existing map projection, and are to this day rarely used in cartographic practice.

4.1.1   Changing the position of the standard line(s)

In section 1.4.1 the possibility was mentioned of replacing the line of zero distortion of the cylindrical and conical projections by two lines along which the nominal scale is preserved. Similarly the point of zero distortion of the azimuthal projection can be replaced by a circle of zero distortion. This type of modification is definitely the most common of all. It is mathematically simple, special distortion properties (equidistance, equal-area, conformality) are easy to maintain, and overall distortion can be substantially reduced by a proper choice of the position of the standard lines. To illustrate this, the redistribution of local scale factors through a change of standard lines will be briefly examined for the normal aspect of the cylindrical equal-area projection. Developing a normal cylindrical projection with two lines of zero distortion instead of one is accomplished by choosing two parallels symmetrical about the equator ($\pm \phi_0$) and representing them in correct length. The x-coordinate of the projection then becomes:

$$x = R \cos \phi_0 \lambda \tag{4.1}$$

The equal-area condition $\sigma = 1$ (equation (1.21)), together with equations (1.8), and the fact that $\partial x/\partial \phi$ and $\partial y/\partial \lambda$ are both zero for cylindrical projections, lead to:

$$\frac{dx}{d\lambda} \frac{dy}{d\phi} = R^2 \cos \phi \tag{4.2}$$

and with the help of equation (4.1):

$$dy = \frac{R}{\cos \phi_0} \cos \phi d\phi \tag{4.3}$$

which becomes after integration (assuming $y = 0$ on the equator):

$$y = \frac{R}{\cos \phi_0} \sin \phi \tag{4.4}$$

Equations (4.1) and (4.4) define the rectangular coordinates for the cylindrical equal-area projection with two standard parallels. The scale factors along the meridians h and the parallels k are given by:

$$h = \frac{dy}{Rd\phi} = \frac{\cos\phi}{\cos\phi_0} \tag{4.5}$$

$$k = \frac{dx}{R\cos\phi d\lambda} = \frac{\cos\phi_0}{\cos\phi} \tag{4.6}$$

As can be seen the product of h and k equals one in every point on the map, which is a consequence of the equal-area property. No distortion occurs along the standard parallels ($\pm\phi_0$). Between the two standard parallels ($|\phi| < |\phi_0|$), the scale factor has a maximum value along the meridians, and a minimum value along the parallels. Between the standard parallels and the edges of the map ($|\phi| > |\phi_0|$) the directions of maximum and minimum scale factor value are reversed. Hence equatorial areas are stretched in the N–S direction, polar areas in the E–W direction. Changing the value of ϕ_0 leads to a redistribution of local scale factors, and a change of the ratio between the length of the equator and the length of the meridians. For $\phi_0 = 0$ there is only one standard line coinciding with the equator, and the stretching of the parallels strongly increases from the equator towards the poles (see Lambert's cylindrical equal-area projection, Canters and Decleir, 1989, p. 159). A proper choice of ϕ_0 leads to a more balanced pattern of distortion, with less extreme scale factors near the edges of the map. Behrmann (1909), who was one of the first to apply a quantitative analysis of distortion to optimise the position of the standard lines of a map projection, obtained a minimum mean angular distortion for the cylindrical equal-area projection with standard parallels at $\pm30°$ (see section 2.2). Repeating Behrmann's analysis for different values of ϕ_0 indeed shows that the position of the standard parallels has a significant impact on overall distortion (figure 4.1).

While only shown here for the cylindrical equal-area projection, similar generalised transformation formulas can be derived for azimuthal and conical equal-area projections, as well as for equidistant and conformal versions of the three conic projection classes. For conformal projections the introduction of two standard lines instead of one is very easy to accomplish. Since conformality implies that scale factors are the same in all directions (see section 1.2.2), it suffices to multiply the x- and y-coordinates of the conformal projection with one standard parallel (or standard point) by a constant scale factor $k_0 < 1$ to obtain the generalised version of the projection. This type of modification is commonly applied in topographic mapping. Generalised azimuthal, cylindrical or conical projections, obtained by introducing one standard circle (azimuthal projections), two standard lines (cylindrical projections), or two standard circular arcs (conical projections), are often called *secant* projections, because similar distortion patterns result from perspective projection of the globe upon a plane, a cylinder or a cone that intersects the spherical surface. The use of the term, however, is regrettable since the principle is applied in a much more general context.

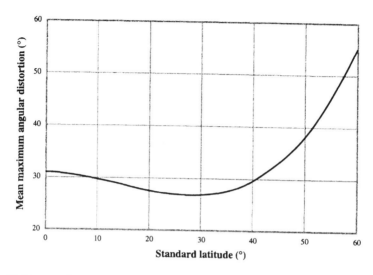

Figure 4.1 Mean maximum angular distortion as a function of the standard latitude for the cylindrical equal-area projection.

4.1.2 Adjusting the geometric qualities of standard map projections

While projections of the conic type have a strict geometric definition with only limited options for modification of the graticule, more general projection classes, especially the pseudocylindrical and the polyconic class, offer many possibilities for the development of generalised map projections that allow independent adjustment of various graticule characteristics. Although several authors have proposed methods for the development of projections with adjustable geometry (e.g. Baar, 1947; McBryde and Thomas, 1949), we will concentrate on a transformation technique that was first suggested by the German cartographer Karlheinz Wagner in 1932, and that is commonly known in the German literature as *Das Umbeziffern*. The technique was formalised, expanded and presented as a general theory for map projection transformation by Karl Siemon (1936, 1937, 1938), and successively applied in various papers and textbooks by its originator (Wagner, 1941, 1944, 1949, 1962, 1982). It deserves special attention because of its general applicability, and its use in the selection procedure that will be proposed in chapter 6.

The transformation method is based on a very simple idea, but provides a powerful mechanism for the development of new map projections. First a well-chosen part of the graticule of an existing projection, bounded by an upper and a lower parallel, and a left and right meridian, is selected. The entire area to be represented is mapped onto this part of the graticule by redefinition of the longitude and latitude value of each meridian and parallel (*Umbeziffern*). Then the graticule is enlarged to the original scale of the parent projection. Restoration of the original scale may be followed by an affine transformation in the x- and y-direction. This permits control of the ratio of the axes of the projection. In *Kartographische Netzentwürfe* (1949), Wagner presented three different transformation methods, all based on Siemon's general theory, and all three preserving some property of the original graticule. He applied them to various types of map projections, and presented several new graticules. Each transformation method will be briefly described below. Some examples of generalised projections with straight as well as curved parallels, that will be used later in this study, will be discussed in detail. For other examples the reader is referred to Wagner's textbook (1949; 2nd edition 1962).

Wagner's first transformation method

Let the original graticule be defined as:

$$x = f_1(u,v)$$
$$y = f_2(u,v)$$

(4.7)

where u and v correspond with the geographical coordinates on the globe. Hence :

$$u = \phi$$
$$v = \lambda$$

(4.8)

For the derived graticule the transformation formulas can be written as (Wagner, 1962):

$$x = Af_1(u,v)$$
$$y = Bf_2(u,v)$$

(4.9)

where u and v are now some function of ϕ and λ:

$$u = u(\phi)$$
$$v = v(\lambda)$$

(4.10)

In principle the functions $u(\phi)$ and $v(\lambda)$ can be chosen freely. However, if certain characteristics of the original graticule are to be maintained, some constraints have to be imposed. Wagner presented three different transformation methods. The first one is a simple linear transformation that guarantees proportional scaling, with $u = m\phi$ and $v = n\lambda$:

$$x = \frac{k_1}{\sqrt{mn}} f_1(m\phi, n\lambda)$$

$$y = \frac{1}{k_2\sqrt{mn}} f_2(m\phi, n\lambda)$$

(4.11)

This transformation will represent the entire world within the part of the parent projection bounded by latitudes $m\pi/2$ and $-m\pi/2$ and longitudes $n\pi$ and $-n\pi$. The pole will be represented by a line, unless the original graticule shows the pole as a point and $m = 1$. The shape of the pole line will depend on the representation of the parallels in the parent projection. If the original graticule has equally spaced parallels and/or meridians the equal spacing will be maintained. By dividing the x- and y-coordinates by the geometric mean of m and n the scale of the original graticule is somewhat restored. The additional parameters k_1 and k_2 allow differential stretching in the direction of the coordinate axes. For example, let us consider Apianus' pseudocylindrical projection with equally spaced parallels and elliptical meridians (Canters and Decleir, 1989, p. 75). The transformation formulas for this projection can be written as:

$$x = R\lambda \cos \psi$$

$$y = R\frac{\pi}{2}\sin \psi \tag{4.12}$$

$$\sin \psi = \frac{2\phi}{\pi}$$

Applying Wagner's first transformation method (equation (4.11)) the following generalised equations are obtained:

$$x = R\frac{k_1}{\sqrt{mn}}n\lambda \cos \psi$$

$$y = R\frac{1}{k_2\sqrt{mn}}\frac{\pi}{2}\sin \psi \tag{4.13}$$

$$\sin \psi = \frac{2m\phi}{\pi}$$

Equations (4.13) define a pseudocylindrical projection with equally spaced parallels, elliptical meridians and pole line. The geometry of the graticule is determined by the four parameters m, n, k_1, k_2. Due to the simple geometry of the pseudocylindrical projection the system of equations is over-defined. All possible graticules defined by (4.13) can be derived by changing the values of three of the four parameters independently and fixing the value of the fourth. Putting $k_2 = 1$ it is easy to show that:

$$m = \sqrt{1 - c^2} \tag{4.14}$$

$$n = \frac{m}{2k_1 p} \tag{4.15}$$

where c is the length of the pole line divided by the length of the equator, and p is the ratio of the axes (central meridian:equator). Wagner (1932) proposed a projection with a pole line half the length of the equator and a correct ratio of the axes by putting $c = 0.5$, $p = 0.5$, and $k_1 = k_2 = 1$ (figure 4.2). This choice of parameters also guarantees that the equator is shown in correct length. The projection is referred to as Wagner VI. Any other parallel can be represented correctly by choosing an appropriate value for $k_1 < 1$, and leaving the other parameters of the projection m, n, and k_2 unchanged. This modification will compress the projection in one direction and will destroy the correct ratio of the axes. The relative length of the pole line will not be altered.

Wagner's second transformation method

The second transformation proposed by Wagner maintains the area distortion pattern of the parent projection. Assuming that $u = 0$ for $\phi = 0$ and $v = 0$ for $\lambda = 0$ it can be shown that this condition is fulfilled when

$$v = n\lambda \tag{4.16}$$

and

$$\sin u = m \sin \phi \tag{4.17}$$

with

$$m = \frac{1}{nAB} \tag{4.18}$$

Restoring the scale of the parent projection as in the previous transformation, and introducing an additional scale factor k for direction dependent stretching, the general formulas for Wagner's equal-area transformation become:

$$x = \frac{k}{\sqrt{mn}} f_1(u, v)$$
$$\tag{4.19}$$
$$y = \frac{1}{k\sqrt{mn}} f_2(u, v)$$

with u and v given by (4.16) and (4.17) respectively. Note that a stretching in the direction of the x-axis k is now attended with an equal compression $1/k$ in the y-direction. This is a necessary condition to maintain the equal-area property of the transformation (see equation (4.18)). The reader who is interested in the full mathematical derivation of the transformation is referred to Wagner (1962, p. 235). Although the transformation can be applied to any projection it is preferably applied to an equal-area graticule since this will produce a generalised equal-area projection. If we consider, for example, Sanson's well-known sinusoidal projection (figure 2.14c), which is defined by:

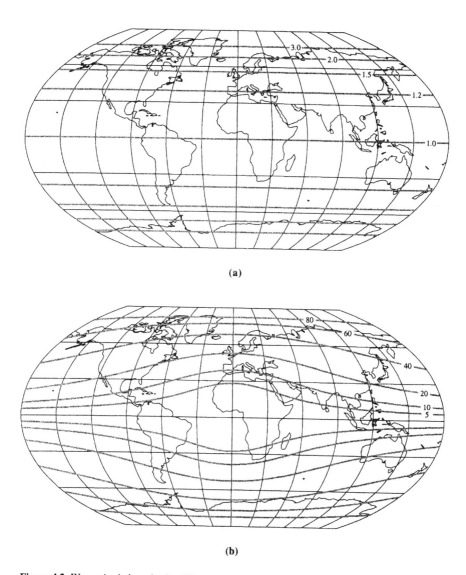

Figure 4.2 Wagner's sixth projection (Wagner VI) with lines of constant area scale (a), and lines of constant maximum angular distortion (b).

$$x = R\lambda \cos \phi$$
$$y = R\phi \tag{4.20}$$

then the application of Wagner's equal-area transformation yields the following general projection equations:

$$x = R \frac{k}{\sqrt{mn}} n\lambda \sqrt{1 - m^2 \sin^2 \phi}$$

$$y = R \frac{1}{k\sqrt{mn}} \arcsin(m \sin \phi)$$

(4.21)

Again the system of equations is over-defined for pseudocylindrical projections. Only two of the three parameters m, n, and k are independent. It is easy to show that for $k = 1$ the parameter m is again given by:

$$m = \sqrt{1 - c^2}$$

(4.14)

with c the length of the pole line divided by the length of the equator. The parameter n is then given by:

$$n = \frac{\arcsin m}{p\pi}$$

(4.22)

with p the ratio of the axes (central meridian:equator). Wagner (1932) presented a special case of the general projection equations (4.21) by putting $c = 0.5$, $p = 0.5$, and $k = 1$. The resulting graticule has a pole line half the length of the equator, a correct ratio of the axes and is, by definition, equal-area (figure 4.3). It is known as Wagner's first projection (Wagner I). The projection was independently re-derived by Kavrayskiy in 1936, and is therefore also known as Kavrayskiy VI (Snyder, 1977). Like all pseudocylindrical equal-area projections with a pole line half the length of the equator, Wagner's first projection shows a strong variation in scale along the central meridian (see also section 2.5.2). Both Wagner's first and Wagner's second transformation (see above) are included in the selection

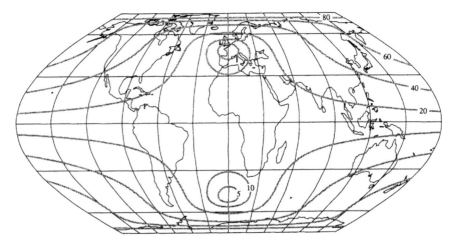

Figure 4.3 Wagner's first projection (Wagner I) with lines of constant maximum angular distortion.

procedure that is presented in chapter 6, to allow the derivation of pseudocylindrical and polyconic map projections with controllable geometric features.

Wagner's third transformation method

To reduce the variation in scale along the central meridian, Wagner proposed a generalised version of his equal-area transformation by introducing an additional parameter m_2 in equation (4.17):

$$\sin u = m_1 \sin(m_2 \phi)$$ (4.23)

and taking this parameter into account in restoring the scale of the graticule:

$$x = \frac{k}{\sqrt{m_1 m_2 n}} f_1(u, v)$$

$$y = \frac{1}{k\sqrt{m_1 m_2 n}} f_2(u, v)$$ (4.24)

Applied to an equal-area graticule, the transformation produces a projection system with adjustable areal scale factor σ:

$$\sigma = \frac{\cos(m_2 \phi)}{\cos \phi}$$ (4.25)

The amount of area distortion can be controlled by specifying a value σ_1 for a chosen parallel ϕ_1. The value of the parameter m_2 is then given by:

$$m_2 = \frac{\arccos(\sigma_1 \cos \phi_1)}{\phi_1}$$ (4.26)

Wagner used his third transformation to develop projections with an area scale $\sigma_1 = 1.2$ on the parallel of 60°, which means that on these projections most of the populated world is represented with less than 20% distortion of area. One of Wagner's projections is obtained by applying the transformation to Sanson's sinusoidal projection, which yields the following general transformation formulas:

$$x = R \frac{k}{\sqrt{m_1 m_2 n}} n\lambda \sqrt{1 - m_1^2 \sin^2(m_2 \phi)}$$

$$y = R \frac{1}{k\sqrt{m_1 m_2 n}} \arcsin[m_1 \sin(m_2 \phi)]$$ (4.27)

For $k = 1$ the graticule can be adjusted using the following relationships:

$$m_1 = \frac{\sqrt{1-c^2}}{\sin\left(m_2 \, \pi/2\right)} \tag{4.28}$$

$$n = \frac{\arcsin\left(\sqrt{1-c^2}\right)}{p\pi} \tag{4.29}$$

where c is again the length of the pole line divided by the length of the equator, p is the ratio of the axes of the projection (central meridian:equator), and m_2 is the parameter which controls the amount of area distortion (equation (4.26)). Putting $c = 0.5$, $p = 0.5$, $k = 1$, and specifying an area distortion of 20% on the equator of 60°, Wagner obtained his second projection (Wagner II), which has much less scale variation along the central meridian than his first (Wagner I) (Wagner, 1949) (figure 4.4).

Generalised projections with curved parallels

Although in the examples above the parent projection is always of the pseudocylindrical type, Wagner applied his transformation also to members of other projection classes. Especially important to this study are Wagner's transformations of transverse azimuthal projections of one hemisphere, since these transformations produce doubly symmetric graticules with curved parallels and curved meridians that are suited for world maps. Three types of transformations proposed by Wagner will be briefly discussed, i.e. the proportional transformation of the transverse azimuthal equidistant projection, the equal-area transformation of the transverse azimuthal equal-area projection, and finally the generalisation of the latter, which supports the development of graticules with adjustable distortion of area. The first and the second transformation are used in the procedure for map projection selection that is proposed in chapter 6.

The equations for the transverse azimuthal equidistant projection are defined as follows:

$$x = R\delta' \sin \lambda'$$
$$y = -R\delta' \cos \lambda' \tag{4.30}$$
$$\delta' = \pi/2 - \phi'$$

where δ' is the angular distance from the centre of the projection (located on the equator), and λ', ϕ' are the meta-longitude and meta-latitude defined with respect to this centre (see also section 2.6). The relationship between meta-coordinates and geographical coordinates is found by applying spherical trigonometry:

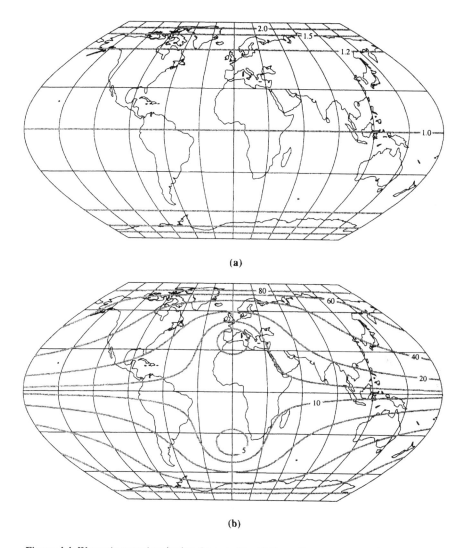

(a)

(b)

Figure 4.4 Wagner's second projection (Wagner II) with lines of constant area scale (a), and lines of constant maximum angular distortion (b).

$$\sin \phi' = \cos \lambda \cos \phi$$

$$\sin \lambda' = \frac{\sin \lambda \cos \phi}{\cos \phi'}$$

$$\cos \lambda' = -\frac{\sin \phi}{\cos \phi'} \qquad (4.31)$$

The meta-latitude is unambiguously determined by the first equation. For the meta-longitude, which ranges from $-\pi$ to π, quadrant adjustment is necessary. The calculation of $\sin\lambda'$ and $\cos\lambda'$ makes identification of the right quadrant possible. Applying Wagner's proportional transformation (equation (4.11)) leads to the following general transformation formulas:

$$x = R\frac{k_1}{\sqrt{mn}}\delta'\sin\lambda'$$

$$y = -R\frac{1}{k_2\sqrt{mn}}\delta'\cos\lambda' \qquad (4.32)$$

$$\delta' = \frac{\pi}{2} - \phi'$$

with

$$\sin\phi' = \cos(n\lambda)\cos(m\phi)$$

$$\sin\lambda' = \frac{\sin(n\lambda)\cos(m\phi)}{\cos\phi'} \qquad (4.33)$$

$$\cos\lambda' = -\frac{\sin(m\phi)}{\cos\phi'}$$

As before the parameter m defines the relative length of the pole line. For $m = 1$ the pole is represented by a point, as in the original projection. For $m < 1$ a pole line of adjustable length is obtained. The parameter n influences the curvature of the parallels. Since on the original projection the curvature of the parallels increases towards the edges of the map, the use of smaller values of n leads to less curved parallels. Once a proper choice of m and n is made, the ratio of the axes can be further controlled by the parameters k_1 and k_2. It is easily found that the ratio between the length of the central meridian and the length of the equator is given by:

$$p = \frac{m}{2k_1k_2n} \qquad (4.34)$$

Since the transverse azimuthal equidistant projection has equally spaced parallels and meridians, all graticules derived from (4.32) and (4.33) have the equator equally divided by the meridians, and the central meridian equally divided by the parallels. Already in 1889, David Aitoff proposed a simple case of Wagner's transformation, which he obtained by stretching the graticule of the transverse azimuthal equidistant projection for one hemisphere in the direction of the x-axis (by a factor 2), and doubling the value of each meridian. Thus the entire world is shown in an ellipse with the longer axis representing the equator and the smaller axis the central meridian (figure 4.5). Using Wagner's transformation method, the projection of Aitoff is obtained by choosing $m = 1$, $n = 0.5$, and $k_1 = k_2 = \sqrt{2}$. Wagner proposed

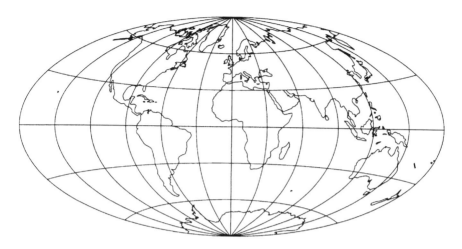

Figure 4.5 Aitoff's projection, obtained by stretching the graticule of the transverse azimuthal equidistant projection for one hemisphere in the direction of the *x*-axis (by a factor 2), and doubling the value of each meridian.

another variant of his transformation by choosing $m = 7/9$ and $n = 5/18$ (Wagner, 1949). This way he obtained a pole line shorter than the equator, and a curvature of the parallels close to the Winkel–Tripel projection (figure 1.9), a graticule which is known for its well-balanced pattern of distortion (see section 2.5.2). A correct ratio of the axes is obtained for $k_1 k_2 = 14/5$ (equation (4.34)). Yet Wagner compressed the graticule horizontally to approximate the Winkel–Tripel projection even better. To obtain an optimal fit he chose $k_1 = 1.4725$ ($0.88\sqrt{14/5}$) and $k_2 = 1.6733$ ($\sqrt{14/5}$) (figure 4.6). The map projection is known as the Aitoff–Wagner (Maling, 1992, p. 244), but is sometimes also referred to as Wagner IX (Snyder, 1993, p. 238).

Wagner also proposed general formulas for the development of equal-area graticules with curved parallels by applying his equal-area transformation (equations (4.16)–(4.19)) to the transverse azimuthal equal-area projection. The equations for the latter are:

$$x = 2R\sin\frac{\delta'}{2}\sin\lambda'$$

$$y = -2R\sin\frac{\delta'}{2}\cos\lambda' \tag{4.35}$$

$$\delta' = \frac{\pi}{2} - \phi'$$

where λ' and ϕ' are the coordinates of the meta-graticule obtained from (4.31). Applying Wagner's equal-area transformation the equations for the generalised projection become:

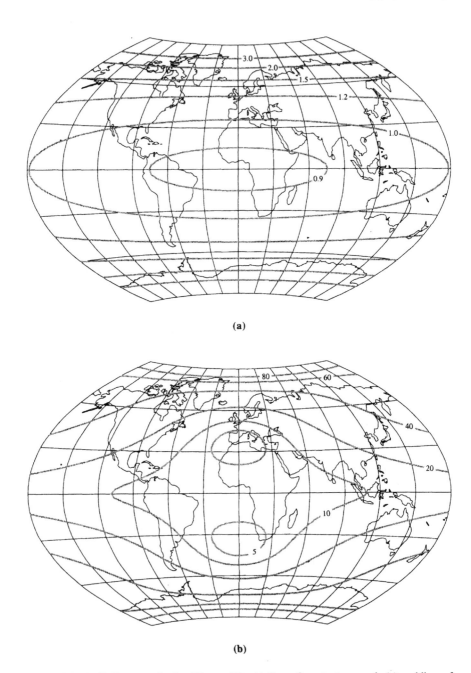

(a)

(b)

Figure 4.6 Aitoff–Wagner projection (Wagner IX) with lines of constant area scale (a), and lines of constant maximum angular distortion (b).

$$x = 2R\frac{k}{\sqrt{mn}}\sin\frac{\delta'}{2}\sin\lambda'$$

$$y = -2R\frac{1}{k\sqrt{mn}}\sin\frac{\delta'}{2}\cos\lambda'$$

(4.36)

with δ' as in (4.35) and:

$$\sin\phi' = \cos(n\lambda)\cos u$$

$$\sin\lambda' = \frac{\sin(n\lambda)\cos u}{\cos\phi'}$$

$$\cos\lambda' = -\frac{\sin u}{\cos\phi'}$$

(4.37)

and with u given by (4.17).

Again the parameters m and n control the length of the pole line and the curvature of the parallels. Once m and n have been chosen the parameter k can be used to adjust the ratio of the axes p, which is given by:

$$p = \frac{1}{k^2}\frac{\sin\left(\dfrac{\arcsin m}{2}\right)}{\sin\left(\dfrac{n\pi}{2}\right)}$$

(4.38)

In 1892, Hammer applied the Aitoff transformation (see above) to the transverse azimuthal equal-area projection, and obtained an equal-area representation of the world in an ellipse with axes in a ratio of 2:1 (figure 4.7) (Hammer, 1892). Hammer's graticule is a special case of Wagner's equal-area projection with curved parallels, obtained by putting $m = 1$, $n = 0.5$ and $k = \sqrt{2}$. More than 40 years later, Eckert repeated Hammer's approach, yet instead of multiplying the x-coordinate of the graticule by a factor 2, and dividing the geographical longitude by the same factor (Aitoff's method of transformation), he used a factor 4 (Eckert–Greifendorff, 1935). This means that in the graticule of the transverse azimuthal equal-area projection the meridian of 45° becomes the meridian of 180°, the meridian of 30° becomes the meridian of 120° and so on, before the multiplication of the x-coordinate by a factor 4 is applied. Using Wagner's transformation, the projection is obtained by putting $m = 1$, $n = 0.25$, and $k = 2$. Due to the small value of n the projection has almost straight parallels, and looks more like a pseudocylindrical projection (see Canters and Decleir, 1989, p. 65). Since in the transverse azimuthal equal-area projection the spacing of the meridians along the equator decreases towards the edges of the map, the ratio of the axes is not correct ($p = 0.4619$).

To avoid the compression of the polar regions, as it occurs on the pointed-polar graticules of Hammer and Eckert, Wagner (1941) proposed a new map projection with a pole line approximately half the length of the equator, slightly curved parallels, and a 2:1 ratio of the axes, which he obtained by representing the

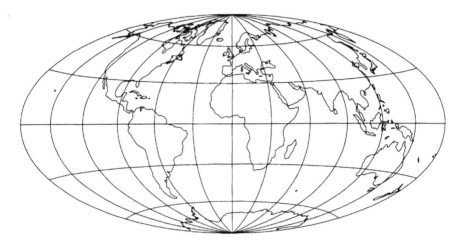

Figure 4.7 Hammer–Aitoff projection, obtained by applying Aitoff's transformation method to the tranverse azimuthal equal-area projection for one hemisphere.

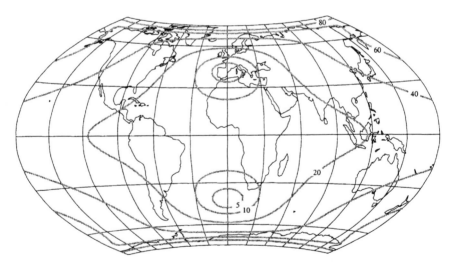

Figure 4.8 Hammer–Wagner projection (Wagner VII) with lines of constant maximum angular distortion.

world within the part of the transverse azimuthal equal-area projection bounded by 65° and –65° in latitude, and 60° and –60° in longitude ($m = \sin 65°$, $n = 1/3$) (figure 4.8). The correct ratio of the axes is obtained by putting $k = 1.4660$ (equation (4.38)). Maling (1992, p. 441) refers to the projection as the

Hammer–Wagner. The projection is the seventh in the sequence of new map projections that are presented in Wagner's *Kartographische Netzentwürfe* (1949). That is why it is also known as Wagner VII.

Finally, Wagner also applied his transformation with adjustable distortion of area, defined by equations (4.16), (4.23), and (4.24), to the transverse azimuthal equal-area projection (Wagner, 1949). The resulting transformation formulas are a generalisation of the equal-area transformation, with the additional parameter m_2 again controlling the amount of area distortion (see equations (4.25)–(4.26)):

$$x = 2R \frac{k}{\sqrt{m_1 m_2 n}} \sin \frac{\delta'}{2} \sin \lambda'$$

$$y = -2R \frac{1}{k\sqrt{m_1 m_2 n}} \sin \frac{\delta'}{2} \cos \lambda' \qquad (4.39)$$

$$\delta' = \pi/2 - \phi'$$

with λ' and ϕ' given by (4.37), and with u given by (4.23).

Wagner again proposed a projection with the world represented within the part of the transverse azimuthal equal-area projection bounded by –65° and 65° in latitude, and –60° and 60° in longitude. Just like for his pseudocylindrical projections with adjustable distortion of area, he specified a maximum distortion of 20% on the parallel of 60°. Hence $m_2 = 0.8855$ (equation (4.26)), $m_1 = 0.9212$ (equation (4.23)), and $n = 1/3$. The ratio of the axes p is given by:

$$p = \frac{1}{k^2} \frac{\sin\left(\dfrac{\arcsin[m_1 \sin(m_2 \pi / 2)]}{2}\right)}{\sin\left(\dfrac{n\pi}{2}\right)} \qquad (4.40)$$

From equation (4.40), the value of k that is required to obtain a correct ratio of the axes can be calculated. Since the present projection and the Hammer–Wagner use the same part of the transverse azimuthal equal-area projection, and represent the equator and the central meridian in a ratio of 2:1, the two have an identical outline and the same value for k (1.4660). Only the size of both graticules and the relative spacing of the parallels are different (figure 4.9). Wagner's projection with curved parallels, pole line, and prescribed area distortion has not been as popular as the Hammer–Wagner. It is known as Wagner VIII, referring to the order in which Wagner listed his nine generalised formulas for world map projections.

4.1.3   Deriving new graticules from polynomial type map projection equations

Although Wagner's transformation method allows independent modification of several graticule characteristics, the nature of the meridians and the parallels is

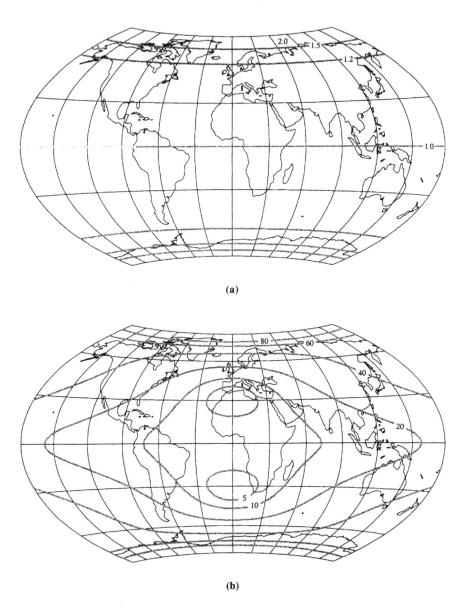

(a)

(b)

Figure 4.9 Wagner's eight projection (Wagner VIII) with lines of constant area scale (a), and lines
of constant maximum angular distortion (b).

strongly determined by the graticule of the parent projection, of which a selected
part is used. Once a number of general graticule characteristics have been defined
(e.g. length of the pole line, ratio of the axes), the appearance of the graticule

cannot be further modified. More general transformations, which provide an almost unlimited capacity for graticule adjustment, are obtained by expressing the relation between map projection coordinates and geographical longitude and latitude by means of power series.

Short review of previous work

A pioneer in the use of power series for map projection development was N.A. Urmayev, who is one of the most important representatives of the Russian school of mathematical cartography. According to Bugayevskiy and Snyder (1995, p. 55), Urmayev was the first to formulate the inverse problem of mathematical cartography, i.e. writing equations of projections from selected distortion values for a number of locations, as opposed to the more common procedure of defining a number of geometric conditions and special distortion properties from which map projection equations are derived, and then calculating distortion values for selected graticule intersections, the so-called direct problem of mathematical cartography (Bugayevskiy and Snyder, 1995, pp. 171–3). Urmayev developed formulas for a general cylindrical projection that allow the construction of an infinite number of graticules with arbitrary distortion characteristics, by representing the scale factor along the meridians h, which for cylindrical projections is a function of the latitude only, by the following polynomial, assuming a unit radius for the generating globe:

$$h = a_0 + a_2\phi^2 + a_4\phi^4 + \dots \tag{4.41}$$

where a_0, a_2, a_4, ... are coefficients that can be determined by solving a set of equations, obtained by specifying the scale factor h for different latitudes. The use of even powers of the latitude only in (4.41) guarantees that the distortion pattern is symmetrical about the equator, which is the case for any regular cylindrical projection. The ordinate y of the resulting graticule is given by:

$$y = \int_0^\phi h\,d\phi \tag{4.42}$$

which after integration gives:

$$y = a_0\phi + \frac{a_2}{3}\phi^3 + \frac{a_4}{5}\phi^5 + \dots \tag{4.43}$$

By defining the above transformation, Urmayev widened the concept of the cylindrical projection. In contrast with the regular cylindrical projection, where angular distortion is zero along the parallel(s) that is (are) represented in correct length, Urmayev developed a cylindrical projection with no angular distortion along four parallels (±20°, ±65°), which is known as Urmayev III (Maling, 1960). The projection has no standard lines, since the parallel that is represented in correct length (the equator) is not free from angular and area distortion (for an illustration of the projection, see Canters and Decleir, 1989, p. 175). Similar projections with other distortion characteristics have been proposed by Pavlov, and by Kharchenko and Shabanova (Maling, 1960; Snyder, 1993, p. 179).

Another example of the use of power series for the development of new map projections is the work by Ginzburg, who proposed five polyconic projections for world maps in the period between 1949 and 1966 that are all symmetric about the equator (four of them are also symmetric about the central meridian), and that are known as the TsNIIGAiK polyconic series, referring to the agency the author was affiliated with (Snyder, 1993, pp. 248–50). The method of development of these and other projections is described in detail in a technical report by Ginzburg and Salmanova (1962, in Russian), and involves two stages (see also Bugayevskiy and Snyder, 1995, pp. 143–9). First a preliminary sketch of the graticule is made which fulfils all geometric properties of the projection to be developed. Next approximate values for the scale factors at different points on the graticule are determined by graphical means or by numerical analysis. The sketch is then corrected until a favourable pattern of distortion is obtained. Since the parallels on Ginzburg's projections are shown as non-concentric circular arcs, each of them equally divided by the meridians, three separate power series can be used to define the graticule, i.e. one for the y-coordinate along the central meridian:

$$y = a_1 + a_3\phi^3 + a_5\phi^5 + \dots \tag{4.44}$$

and two for the x and y-coordinates along the outer meridian:

$$x = b_0 + b_2\phi^2 + b_4\phi^4 + \dots \tag{4.45}$$

$$y = c_1 + c_3\phi^3 + c_5\phi^5 + \dots \tag{4.46}$$

The use of odd powers of the latitude in (4.44) and (4.46), and of even powers of the latitude in (4.45), guarantees that the graticule is symmetrical about the equator. From the coordinates for a set of points lying on the central and outer meridian, and taken from the sketch, polynomial coefficients in (4.44), (4.45), and (4.46) can be derived by least-squares approximation. Once the coefficients are known, other graticule intersections can be determined by interpolation. Boginskiy (1972) applied the same technique for the development of new projections with specified distortion patterns, making use of specific forms of general algebraic polynomials, each one reflecting different geometric conditions (see below).

Some authors have also proposed the use of polynomial equations for the development of new projections, optimising some local or global measure of distortion. Baetslé (1970), for example, proposed general formulas for the azimuthal projection by defining the radial distance from each point on the map to the centre of the projection r by means of the following polynomial:

$$r = \delta + a_3\delta^3 + 0^5 \tag{4.47}$$

where δ stands for the angular distance from the centre of the projection on the globe, and 0^5 represents all terms of fifth-order and higher. He then derived approximate values of a_3 for well-known azimuthal projections, and presented various other solutions minimising different local distortion measures. Snyder (1985) presented polynomial formulas for pseudocylindrical equal-area projections with a pole line, and for pseudocylindrical equal-area projections that show the pole as a point. Both sets of formulas are based on the following general equations for pseudocylindrical equal-area projections:

$$x = R^2 \lambda \cos \phi \frac{1}{dy \Big/ d\phi}$$

$$y = Rf(\phi)$$

(4.48)

where $f(\phi)$ and $dy/d\phi$ are approximated by appropriate power series for the flat-polar and the pointed-polar case. He determined optimal values for the polynomial coefficients by minimising overall distortion, applying Airy's well-known minimum-error criterion (equation (2.5)). Since for flat-polar projections distortion is infinite at the poles, the range of latitude used in minimising the distortion was limited to $\pm 75°$ (Snyder, 1985, pp. 120–31).

Instead of defining polynomial equations for traditional classes of projections, the use of polynomials also allows the definition of more general types of projections with a more complex graticule geometry. In its most general form, assuming a unit radius for the generating globe, the relationship between the coordinates in the map plane and the coordinates on the globe can be expressed by the following two polynomials:

$$x = \sum_{i=0}^{n} \sum_{j=0}^{n-i} C_{ij} \lambda^i \phi^j$$

(4.49)

$$y = \sum_{i=0}^{n} \sum_{j=0}^{n-i} C'_{ij} \lambda^i \phi^j$$

(4.50)

with x,y the map projection coordinates, λ and ϕ the geographical longitude and latitude, and C_{ij} and C'_{ij} the polynomial coefficients defining the properties of the graticule. The order of the polynomials n will determine the flexibility of the transformation. By putting appropriate constraints upon the values of the polynomial coefficients a multitude of map projection graticules can be derived, varying from graticules with a very complex, irregular geometry, to graticules that belong to traditional map projection classes. Applying this strategy, the present author derived various new map projections with arbitrary distortion characteristics (neither conformal, nor equal-area) that are suited for global mapping purposes (Canters, 1989). The graticules were obtained by minimising the distortion of finite distance, using the global distortion measure defined in the previous chapter (see sections 3.4–3.6). They will be discussed at length in chapter 5, which is entirely devoted to the development of new map projections with low finite distortion. Following this author's work, Laskowksi (1991) presented a general polynomial type polyconic map projection with low error (the so-called *Tri-*

Optimal projection) by minimising a mixed local-global distortion measure that takes into account both infinitesimal and finite components of distortion.

Definition of geometric constraints

We will now investigate how different geometric conditions that are important in the definition of a new map projection can be imposed upon a set of mapping equations of the polynomial type. Geometric properties may refer to the level of symmetry of the graticule, the nature and spacing of the parallels and the meridians, the ratio of the axes, the representation of the pole, ... For a full discussion of the role of graticule geometry in the selection of a map projection the reader is referred to chapter 6. As we will see, most geometric properties that are of practical use for mapping can be obtained by putting some of the polynomial coefficients in the general mapping equations equal to zero, others by defining a mathematical relationship between two or more coefficients. In both cases the number of coefficients to be chosen or optimised will be reduced. The following discussion will focus on the definition of polynomial equations for map projections with arbitrary distortion characteristics, i.e. map projections that are neither conformal, nor equal-area. Conformal and equal-area transformations will be addressed in section 4.2. Map projection equations will be written out for fifth-order polynomials, as these will be used for the development of new projections in the next chapter.

Symmetry conditions

Starting with two general fifth-order polynomials (equations (4.49) and (4.50)), and letting the origin of the map coordinates coincide with the point of intersection of the equator and the zero meridian, C_{00} and C'_{00} will both become zero, and the transformation will have as much as 40 (!) independent coefficients. A general transformation of this type will allow the development of highly irregular graticules that may be optimally adapted to a given area of interest by an appropriate choice of coefficient values. However, the number of coefficients can be substantially reduced by imposing simple symmetry conditions that are common to most small-scale map projections. For example, if the coefficients of even powers of λ in (4.49), and of odd powers of λ in (4.50) are both made zero the transformation will become symmetrical about the y-axis. Applying this constraint reduces the number of coefficients in the optimisation problem from 40 to 20:

$$x = C_{10}\lambda + C_{11}\lambda\phi + C_{30}\lambda^3 + C_{12}\lambda\phi^2 + C_{31}\lambda^3\phi + C_{13}\lambda\phi^3 + C_{50}\lambda^5$$
$$+ C_{32}\lambda^3\phi^2 + C_{14}\lambda\phi^4 \tag{4.51}$$

$$y = C'_{01}\phi + C'_{20}\lambda^2 + C'_{02}\phi^2 + C'_{21}\lambda^2\phi + C'_{03}\phi^3 + C'_{40}\lambda^4 + C'_{22}\lambda^2\phi^2$$
$$+ C'_{04}\phi^4 + C'_{41}\lambda^4\phi + C'_{23}\lambda^2\phi^3 + C'_{05}\phi^5 \tag{4.52}$$

Symmetry about the x-axis is obtained by putting all odd powers of ϕ in (4.49), and all even powers of ϕ in (4.50) equal to zero. Most projections that are used for global mapping purposes are symmetrical about the equator and the central meridian in the normal aspect. If projections of this type are to be developed, equations (4.49) and (4.50) will reduce to:

$$x = C_{10}\lambda + C_{30}\lambda^3 + C_{12}\lambda\phi^2 + C_{50}\lambda^5 + C_{32}\lambda^3\phi^2 + C_{14}\lambda\phi^4 \tag{4.53}$$

$$y = C'_{01}\phi + C'_{21}\lambda^2\phi + C'_{03}\phi^3 + C'_{41}\lambda^4\phi + C'_{23}\lambda^2\phi^3 + C'_{05}\phi^5 \tag{4.54}$$

Spacing of the parallels and the meridians

Other geometric constraints that may be very useful, and that can easily be imposed are the maintenance of equal spacing of the parallels along the straight central meridian and, for world maps, the maintenance of equal spacing of the meridians along the straight equator. The first condition is satisfied by making the y-coordinate a linear function of ϕ for $\lambda = 0$, given that $x = 0$ for $\lambda = 0$. For graticules that are symmetric about the central meridian the x-coordinate will still be defined by (4.51), yet the y-coordinate will become:

$$y = C'_{01}\phi + C'_{20}\lambda^2 + C'_{21}\lambda^2\phi + C'_{40}\lambda^4 + C'_{22}\lambda^2\phi^2 + C'_{41}\lambda^4\phi + C'_{23}\lambda^2\phi^3 \tag{4.55}$$

To obtain a straight equator that is evenly divided by the meridians the x-coordinate must be a linear function of λ for $\phi = 0$, given that $y = 0$ for $\phi = 0$. Many doubly symmetric projections that are used for world maps have meridians and parallels that are both equally spaced along the axes of symmetry. General polynomial equations for this type of projection are defined as follows:

$$x = C_{10}\lambda + C_{12}\lambda\phi^2 + C_{32}\lambda^3\phi^2 + C_{14}\lambda\phi^4 \tag{4.56}$$

$$y = C'_{01}\phi + C'_{21}\lambda^2\phi + C'_{41}\lambda^4\phi + C'_{23}\lambda^2\phi^3 \tag{4.57}$$

The equal spacing of the parallels and meridians can easily be extended to the maintenance of equidistance along central meridian and equator by simply putting C'_{01} in (4.55) and (4.57), and C_{10} in (4.56) equal to one.

Ratio of the axes

One geometric constraint that is especially important for the development of doubly symmetric world map projections is the maintenance of the correct ratio of the axes. The way this constraint is imposed will depend on the spacing of the parallels and meridians. If the equator and the central meridian are both equally divided (see equations (4.56) and (4.57)) then the correct ratio of the axes is obtained by putting $C_{10}=C'_{01}$. If both are not equally divided, the dependence between the polynomial coefficients that is required to obtain the correct ratio can be derived from (4.53) and (4.54). A correct ratio of the axes implies that:

$$x_E = 2y_P \tag{4.58}$$

with x_E the maximum value of (4.53) on the equator ($\phi = 0$), and y_P the maximum value of (4.54) on the central meridian ($\lambda = 0$). Hence it follows that:

$$C_{10}\pi + C_{30}\pi^3 + C_{50}\pi^5 = 2\left(C'_{01}\frac{\pi}{2} + C'_{03}\frac{\pi^3}{8} + C'_{05}\frac{\pi^5}{32} \right) \tag{4.59}$$

Rearranging the terms one gets:

$$C_{10} = C'_{01} + (C'_{03} - 4C_{30})\frac{\pi^2}{4} + (C'_{05} - 16C_{50})\frac{\pi^4}{16} \tag{4.60}$$

As can be seen, imposing the 2:1 ratio of the axes diminishes the number of coefficients to be chosen freely by one. When applying (4.60), the coefficient C_{10} is used as an adjustment factor to restore the correct ratio of the axes, its value depending entirely on the value of the other coefficients. If the central meridian is equally divided, and the equator is not, or vice versa, then (4.60) will respectively simplify to:

$$C_{10} = C'_{01} - 4C_{30}\frac{\pi^2}{4} - 16C_{50}\frac{\pi^4}{16} \tag{4.61}$$

or

$$C'_{01} = C_{10} - C'_{03}\frac{\pi^2}{4} - C'_{05}\frac{\pi^4}{16} \tag{4.62}$$

Note that in (4.62) C'_{01} has been chosen as the dependent coefficient instead of C_{10}, this to permit the value of C_{10} to be set equal to one in case one should want to obtain true scale along the equator.

Nature of the parallels and the meridians

Since in (4.49) and (4.50) both x and y are a function of the longitude and the latitude, graticules will have curved parallels and meridians. If one wants to develop a graticule with straight parallels, straight meridians, or both, the functional relationships between the coordinates in the map and the geographical coordinates on the globe have to be defined otherwise. Graticules with straight parallels can be developed by making the y-coordinate a function of the latitude only. Equation (4.50) then reduces to:

$$y = \sum_{i=0}^{n} C'_{0i}\phi^i \tag{4.63}$$

Similarly a graticule with straight meridians can be obtained by making the x-coordinate a function of the longitude only. Equation (4.49) then becomes:

$$x = \sum_{i=0}^{n} C_{i0}\lambda^i \tag{4.64}$$

A graticule with straight parallels and straight meridians, orthogonal to the parallels, is obtained by applying both (4.63) and (4.64).

Geometric constraints related to symmetry conditions, the spacing of the parallels and the meridians, and the ratio of the axes, as defined above, can also be applied to these more specific transformations, the only difference being that some of the coefficients in the above expressions will vanish. For example, for a doubly symmetric pseudocylindrical projection with equally spaced parallels that are all equally divided by the meridians (a so-called *true* pseudocylindrical projection, see section 1.4.1), (4.56) and (4.57) will reduce to:

$$x = C_{10}\lambda + C_{12}\lambda\phi^2 + C_{14}\lambda\phi^4 \tag{4.65}$$

$$y = C'_{01}\phi \tag{4.66}$$

leaving only four instead of eight coefficients to be optimised.

Length of the pole line

When developing flat-polar projections with straight parallels and twofold symmetry, it may be useful to be in control of the length of the pole line. The most general formulation of the x-coordinate for a doubly symmetric graticule with straight parallels is given by (4.53). Fixing the length of the pole line can be achieved by expressing the maximum value of the x-coordinate along the pole line x_P as a fraction of the maximum value of the x-coordinate along the equator x_E:

$$x_E = kx_p \tag{4.67}$$

From (4.53) it follows that the necessary condition to obtain a graticule with equator and pole line in a ratio of k is given by:

$$\begin{aligned} C_{10}\lambda_E + C_{30}\lambda_E^3 + C_{50}\lambda_E^5 &= k(C_{10}\lambda_P + C_{30}\lambda_P^3 + C_{12}\lambda_P\phi_P^2 + C_{50}\lambda_P^5 \\ &\quad + C_{32}\lambda_P^3\phi_P^2 + C_{14}\lambda_P\phi_P^4) \end{aligned} \tag{4.68}$$

with λ_E and λ_P equal to π and ϕ_P equal to $\pi/2$. By choosing one of the six coefficients in (4.68), and expressing it as a function of the other five, the total number of coefficients to be optimised is reduced by one. For a *true* pseudocylindrical projection, the x-coordinate is given by (4.65), and (4.68) will simplify to:

$$C_{10}(\lambda_E - k\lambda_P) = (C_{12} + C_{14}\phi_P^2)k\lambda_P\phi_P^2 \tag{4.69}$$

4.2   POLYNOMIAL TRANSFORMATION OF MAP PROJECTION COORDINATES

In the previous section various mapping equations of the polynomial type have been defined to transform positions from the curved surface of the Earth onto a flat map. However, power series may also be used to transfer data from one map projection to another. This may be done to reduce computation time if the number of coordinates to be transformed is very large, or to transfer data between maps with incomplete definitions of map projection parameters (Snyder, 1985, p. 15; Maling, 1992, p. 420). Polynomial transformation of map projection coordinates may also be applied to alter the distortion patterns of any standard map projection to obtain less distortion within the area to be mapped. In the last two decades various authors have used complex-algebra polynomials to reduce map error obtained with standard conformal map projections. On the other hand, polynomial transformation of map projection coordinates for the development of arbitrary projections or equal-area projections, with less distortion, has hardly been explored.

In the following we will discuss different types of polynomial transformation in the map plane. Special attention will be paid to the definition of transformations that maintain some of the properties of the original graticule. From a practical point of view these transformations are particularly important. If one chooses a projection with well-defined properties in accordance with the purpose of the map, then it is obvious that these properties should be maintained if an error-reducing transformation is applied. As will be shown in chapter 6, map properties that are important in the selection process may relate to the geometry of the graticule (symmetry conditions, shape of the parallels and the meridians, ...), as well as to special distortion characteristics (conformality, equidistance, maintenance of correct area, ...).

First we will discuss polynomial transformation in its most general form, and see how different geometric constraints can be imposed. Next we will define a polynomial equal-area transformation which is particularly useful for small-scale applications, and which will be extensively used in the next chapter. Finally, we will look at some of the work that has been done with complex-algebra polynomials. Although complex-algebra transformations are less frequently applied in small-scale cartography, they may be of use if conformality is a prime concern.

4.2.1   General polynomial transformations

In its most general form a polynomial transformation in the map plane can be defined as:

$$X = \sum_{i=0}^{n} \sum_{j=0}^{n-i} C_{ij} x^i y^j \tag{4.70}$$

$$Y = \sum_{i=0}^{n} \sum_{j=0}^{n-i} C'_{ij} x^i y^j \qquad (4.71)$$

with x,y the map projection coordinates of the parent projection, X,Y the transformed coordinates, and C_{ij} and C'_{ij} the polynomial coefficients to be defined so that favourable distortion characteristics are obtained.

If the original map projection has some geometric properties that need to be preserved, this can be accomplished by imposing appropriate constraints on the value of the polynomial coefficients. Since equations (4.70) and (4.71) are similar to the general polynomial equations that were defined for the mapping of the surface of the globe onto the plane (equations (4.49) and (4.50)), with x taking the role of λ and y taking the role of ϕ, geometric properties are imposed as before, and the reader is referred to section 4.1.3 for further details. For example, if a graticule that is symmetrical about the equator and the central meridian is to be adjusted by a fifth-order polynomial transformation, without loss of its twofold symmetry, then equations (4.70) and (4.71) have to be reduced to:

$$X = C_{10}x + C_{30}x^3 + C_{12}xy^2 + C_{50}x^5 + C_{32}x^3y^2 + C_{14}xy^4 \qquad (4.72)$$

$$Y = C'_{01} y + C'_{21} x^2y + C'_{03} y^3 + C'_{41} x^4y + C'_{23} x^2y^3 + C'_{05} y^5 \qquad (4.73)$$

If, besides the twofold symmetry, the parallels and the meridians on the original graticule are equally spaced along the axes, then adjustment of the projection without loss of the property of equal spacing can be achieved by applying the following transformation:

$$X = C_{10}x + C_{12}xy^2 + C_{32}x^3y^2 + C_{14}xy^4 \qquad (4.74)$$

$$Y = C'_{01} y + C'_{21} x^2y + C'_{41} x^4y + C'_{23} x^2y^3 \qquad (4.75)$$

A doubly symmetric projection with a correct ratio of the axes is obtained by applying (4.72) and (4.73), thereby assuming that:

$$C_{10}x_E + C_{30}x_E^3 + C_{50}x_E^5 = 2(C'_{01} y_P + C'_{03} y_P^3 + C'_{05} y_P^5) \qquad (4.76)$$

with x_E the maximum x-coordinate on the equator, and y_P the maximum y-coordinate on the central meridian in the original graticule. If the original projection has a 2:1 ratio of the axes, then $x_E = 2y_P$. By substituting $2y_P$ for x_E in (4.76) and rearranging the terms one gets:

$$C_{10} = C'_{01} + (C'_{03} - 4C_{30})y_P^2 + (C'_{05} - 16C_{50})y_P^4 \qquad (4.77)$$

Graticules with straight parallels can be developed by transformation of a cylindrical or a pseudocylindrical projection, making the Y-coordinate of the transformed graticule a function of the y-coordinate of the parent projection only. Equation (4.71) then reduces to:

$$Y = \sum_{i=0}^{n} C'_{0i} y^i \tag{4.78}$$

Similarly a graticule with straight meridians, running parallel with the Y-axis, can be obtained by transforming a cylindrical projection, this time making the X-coordinate a function of the x-coordinate only. Equation (4.70) then becomes:

$$X = \sum_{i=0}^{n} C_{i0} x^i \tag{4.79}$$

A graticule with straight parallels and straight meridians, orthogonal to the parallels, is obtained by transforming a cylindrical projection, applying both (4.78) and (4.79). This is the simplest transformation of all. It merely involves non-linear differential stretching and compression in the X and Y direction and produces what is called a generalised cylindrical projection (Starostin *et al.*, 1981; see also section 1.4.2, table 1.3).

Controlling the length of the pole line for doubly symmetric polynomial transformations that maintain the straightness of the parallels of a pseudocylindrical or cylindrical parent projection, is again achieved by expressing the maximum value of the X-coordinate along the pole line as a fraction of the maximum value of the X-coordinate along the equator. From (4.72) it follows that the necessary condition to obtain a graticule with equator and pole line in a ratio of k is given by:

$$\begin{aligned} C_{10} x_E + C_{30} x_E^3 + C_{50} x_E^5 = {}& k(C_{10} x_P + C_{30} x_P^3 + C_{12} x_P y_P^2 + C_{50} x_P^5 \\ & + C_{32} x_P^3 y_P^2 + C_{14} x_P y_P^4) \end{aligned} \tag{4.80}$$

with x_E the x-coordinate of the end of the equator, and x_P, y_P the x,y-coordinates of the end of the pole line in the original graticule. If all parallels are equally divided by the meridians in the original graticule, and this property is to be maintained in the transformed graticule, then C_{30}, C_{50} and C_{32} in (4.72) have to be made zero and (4.80) will simplify to:

$$C_{10}(x_E - k x_P) = (C_{12} + C_{14} y_P^2) k x_P y_P^2 \tag{4.81}$$

If the ratio between the length of the equator and the length of the pole line of the original graticule is to be maintained, instead of arbitrarily modified, then x_E can be substituted for $k x_P$ and equation (4.81) becomes:

$$C_{12} + C_{14} y_P^2 = 0 \tag{4.82}$$

4.2.2   Equal-area polynomial transformations

Maintenance of the equal-area property requires the definition of polynomial equations that satisfy the general condition for an equal-area transformation in the plane:

$$\frac{\partial X}{\partial x}\frac{\partial Y}{\partial y} - \frac{\partial X}{\partial y}\frac{\partial Y}{\partial x} = 1 \tag{4.83}$$

where $\partial X/\partial x$, $\partial X/\partial y$, $\partial Y/\partial x$, $\partial Y/\partial y$ are the partial derivatives of X and Y with respect to x and y.

From (4.83) two simple solutions can be derived:

$$X = f_1(x)$$
$$Y = \frac{y}{dX/dx} \tag{4.84}$$

and

$$X = \frac{x}{dY/dy}$$
$$Y = f_2(y) \tag{4.85}$$

Transformation (4.84) is symmetric about the X-axis, and preserves the proportional scaling of the parallels along the central meridian, if the latter coincides with the y-axis in the original graticule. Straight meridians, parallel to the Y-axis, remain straight since X is a function of x only (figure 4.10a). Transformation (4.85) is symmetric about the Y-axis, and does not alter the relative spacing along the x-axis. Straight parallels remain straight (figure 4.10b).
By putting

$$f_1(x) = \sum_{i=1}^{n} C_i x^i \tag{4.86}$$

and

$$f_2(y) = \sum_{i=1}^{n} C'_i y^i \tag{4.87}$$

two polynomial transformations are obtained that do not alter the area distortion pattern of the original graticule, each with its own geometric properties and, with n

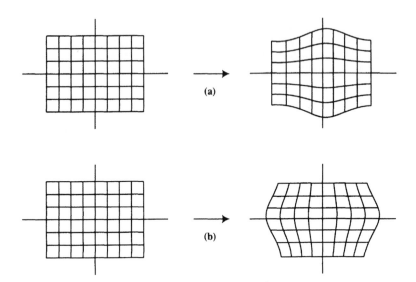

Figure 4.10 Geometric characteristics of the first (a) and the second equal-area polynomial transformation (b).

coefficients to be optimised. A higher level of flexibility is obtained by combining (4.84) and (4.85) into one transformation with $2n$ coefficients:

$$X' = \frac{X}{dY'/dY}$$
$$Y' = f_2(Y)$$

(4.88)

with X and Y as in (4.84) (Canters, 1991). When applied to an equal-area map projection, a new equal-area graticule is obtained. Geometric constraints, as defined in section 4.2.1, are again easy to impose. For examples, the reader is referred to chapter 5 (see sections 5.3.2, 5.4.2, 5.4.3).

If we consider only the first-order term of (4.86), then (4.84) reduces to a simple affine transformation with inverse expansion and compression along both coordinate axes:

$$X = C_1 x$$
$$Y = \frac{1}{C_1} y$$

(4.89)

Applied to Lambert's oblique azimuthal equal-area projection, it changes the circular isocols into ovals, which can be given any arbitrary orientation by a simple rotation of the coordinate axes prior to the transformation. A good choice of the affine coefficient and a proper orientation of the axes will lead to less variation of scale and angular distortion for elongated areas. Tobler (1974) presented examples of this transformation for areas of different size (Eurafrica, United States, Michigan). He demonstrated that the maximum scale factor at the extreme points of the geographical area along both coordinate axes can be made equal to the geometrical average of the maximum scale factors at the same points in the unmodified projection by choosing

$$C_1 = \sqrt{a_1 \Big/ a_2} \qquad\qquad (4.90)$$

where a_1 and a_2 are the maximum scale factors for the extreme points on the original projection. The use of (4.90) offers a simple mechanism to derive an equal-area projection with the region to be mapped approximately enclosed within an oval line of constant scale. Double polynomial transformation using higher-order terms, as suggested above, allows a more flexible adjustment of the shape of the isocols, and thus a better limitation of error for regions of arbitrary shape and size.

It should be noted that consecutive transformation along two perpendicular directions, as in (4.88), is not new. It has also been applied by Snyder (1988) for the development of equal-area projections with low distortion. Starting from an oblique azimuthal equal-area projection centred on the region to be mapped (with circular isocols), Snyder altered the spacing in the x- and y-direction in a similar way as above to produce equal-area map projections with oval, rectangular or rhombic isocols, that have less distortion of angles and scale for non-circular regions. Instead of working with polynomials, Snyder uses sine functions to define the spacing in the x- and y-direction for both transformations. The final formulas for his so-called *oblated equal-area projection series* are:

$$X' = mR\sin(2M/m)\frac{\cos N}{\cos(2N/n)}$$
$$Y' = nR\sin(2N/n) \qquad\qquad (4.91)$$

with

$$M = \arcsin(x/2) \qquad\qquad (4.92)$$

$$N = \arcsin\left(\frac{y}{2}\frac{\cos M}{\cos(2M/m)}\right) \qquad\qquad (4.93)$$

where x,y are the coordinates for the oblique azimuthal equal-area projection, and m, n are constant parameters, which define the shape of the isocols ($n > m$). Isocols are stretched along the y-axis, the ratio of their axes is $n : m$. For $m = 2$, $n = 2$, the original azimuthal projection with circular isocols is obtained. By rotating the coordinate axes of the parent projection before the transformation is applied, the oval, rectangular or rhombic isocols can be oriented along the main directions of the region to be mapped, leading to less variation of scale and angular distortion. Snyder (1988) presented examples with oval isocols of different eccentricity for the mapping of the Atlantic Ocean, as well as a map of the conterminous U.S. with rectangular isocols (figure 4.11, figure 4.12).

Snyder's transformation is less flexible than the double polynomial transformation defined above (equation (4.88)), and only applies to Lambert's azimuthal equal-area projection, yet it has the practical advantage that isocols can be optimally adapted to the region of interest by experimenting with different values for the constants m and n, and different angles of rotation. Polynomial transformations of higher-order allow a more general modelling of the shape of the isocols, yet the value of the coefficients can only be determined by numerical optimisation (see sections 5.1, 5.2).

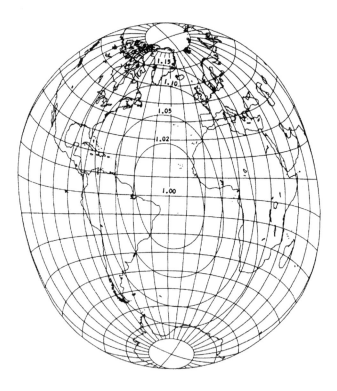

Figure 4.11 Oblated equal-area projection with oval isocols for the mapping of the Atlantic Ocean (from Snyder, 1988, p. 349; reprinted with permission from the American Congress on Surveying and Mapping).

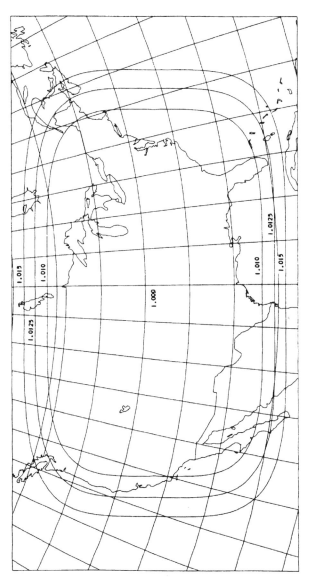

Figure 4.12 Oblated equal-area projection with rectangular isocols for maps of the conterminous U.S. (from Snyder, 1988, p. 352; reprinted with permission from the American Congress on Surveying and Mapping).

4.2.3   Complex-algebra polynomial transformations

For quite some time, the technique of polynomial transformation has been used for the development of conformal map projections with less scale variation. In 1856, Chebyshev stated that a conformal projection has the least possible overall distortion if the region to be mapped is bounded by a line of constant scale (Snyder, 1993, p. 140). Mathematical proof was given by Grave in 1896. Starting from Chebyshev's theorem, several authors have developed low-error graticules by adapting the shape of the isocols of standard map projections to surround the region to be mapped, using complex-algebra polynomial transformations.

The general condition for a conformal transformation is given by the Cauchy–Riemann equations, which state that any map projection that is conformal and represented by a set of isometric coordinates (x,y), is also conformal when transformed to another set of isometric coordinates (X,Y) provided that

$$\frac{\partial X}{\partial x} = \frac{\partial Y}{\partial y} \qquad\qquad \frac{\partial X}{\partial y} = -\frac{\partial Y}{\partial x} \qquad\qquad (4.94)$$

In 1932, Driencourt and Laborde (1932, p. 202) presented the following complex-algebra polynomial transformation fitting the Cauchy–Riemann equations :

$$X + iY = \sum_{j=1}^{n} (A_j + iB_j)(x + iy)^j \qquad\qquad (4.95)$$

where $i^2 = -1$, n is a positive integer, and A_j and B_j are any real coefficients.

Expanding (4.95) and separating the real and imaginary portions:

$$X = A_1 x - B_1 y + A_2 x^2 - 2B_2 xy + A_3 x^3 - 3A_3 xy^2 - 3B_3 x^2 y + B_3 y^3 + \dots \qquad (4.96)$$

$$Y = A_1 y + B_1 x + 2A_2 xy + B_2 x^2 - B_2 y^2 + 3A_3 x^2 y - A_3 y^3 + B_3 x^3 - 3B_3 xy^2 + \dots \qquad (4.97)$$

As can be seen, equations (4.96) and (4.97) are identical to the general equations defined in 4.2.1, except for some dependencies between the coefficients, dictated by the Cauchy–Riemann conditions. Geometric constraints can be defined as above.

Laborde applied the transformation to the transverse Mercator projection of a conformal sphere for the mapping of Madagascar. Choosing $n = 3$ and appropriate values for the coefficients, he obtained a conformal projection with its central line aligned with the oblique direction of the island rather than with the central meridian of the transverse Mercator (Bugayevskiy and Snyder 1995, p. 199). Miller (1953) took a similar approach to develop a low-error conformal map projection for Europe and Africa. Applying (4.95) with $n = 3$, he transformed an oblique stereographic projection of the area, changing the lines of constant scale from circles to ovals with the major axis lying along the central meridian. Miller's transformation is given by

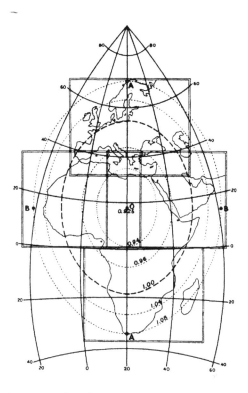

Figure 4.13 Miller's low-error conformal map projection with oval isocols for maps of Europe and Africa (from Miller, p. 407; © 1953 by the American Geographical Society. All rights reserved. Published 1953).

$$X = Kx\left[1 - \frac{Q}{12}(3y^2 - x^2)\right]$$

$$Y = Ky\left[1 + \frac{Q}{12}(3x^2 - y^2)\right]$$

(4.98)

where X and Y are the rectangular coordinates on the new projection, x and y are the rectangular coordinates on an oblique stereographic projection having the same centre and orientation, and K and Q are arbitrarily chosen parameters. Miller derived appropriate values for K and Q (0.9245 and 0.2522 respectively), so that Europe and Africa are surrounded by an oval isocol with a scale ratio ($1/K$) equal to the reciprocal of the scale ratio at the centre of the projection (K) (figure 4.13). It can easily be seen that equations (4.98) are obtained by putting $A_1=1$ and $A_3=Q/12$ in (4.96) and (4.97), making all the other coefficients equal to zero, and introducing the scale conversion factor K. In 1955, Miller adapted the transformation to derive new conformal projections with oblique, oval-shaped isocols for Central Asia and

Australasia (Miller, 1955; see also Sprinsky and Snyder, 1986). Lee (1974) followed the same approach to develop a map with oblique oval isocols for the Pacific Ocean.

While Miller applied a third-order polynomial to obtain projections with oval isocols, other authors have used higher-order transformations to derive map projections with isocols of more irregular shapes. Reilly (1973) used a sixth-order polynomial for the development of a new conformal map projection for the topographic mapping of New Zealand. Starting from the regular Mercator, and applying the method of least squares to determine the value of the coefficients, he developed a graticule with lines of constant scale roughly following the outlines of the two main islands. Snyder (1984a) took a similar approach to develop a new low-error conformal projection for a 50-state map of the United States, using a tenth-order polynomial (figure 4.14). He also proposed simplified forms of higher-order complex polynomials to develop low-error conformal projections with regular-shaped isocols (Snyder, 1984b). González-López (1995) presented a low-error complex-algebra transformation of the transverse Mercator projection ($n = 5$) for the mapping of Chile, and of Lambert's conformal conical projection ($n = 8$) for the mapping of the Mediterranean Sea.

The possibilities of transforming standard conformal projections by means of equation (4.95) are boundless. On the other hand, one should realise that the principle of least squares, applied to a limited set of points, and using a complex polynomial of prescribed order, will not produce the theoretically best solution satisfying Chebyshev's theorem. Indeed, it is not possible to make the isocols follow almost any prescribed pattern and thus to create real *minimum-error* projections for any region of interest. As suggested by Snyder (1985, p. 81) it is therefore more correct to designate projections that are obtained by optimisation of the coefficients in expression (4.95) as *low-error*, instead of *minimum-error* projections.

In the *Oxford Hammond Atlas of the World*, first published in 1993, a new projection is introduced which is known as *Hammond's optimal conformal projection*, and which is said to present "... the optimal view of an area by reducing shifts in scale over an entire region to the minimum degree possible". The projection is used for all continental maps in the atlas. It is obtained by complex-algebra transformation of the azimuthal stereographic projection, yet polynomial coefficients are determined in such a way that one of the lines of constant scale coincides with a highly smoothed version of the outline of the continent to be mapped, thus fulfilling the minimum-error condition for the enclosed region. The curve that delimits the area with minimal distortion is explicitly shown on each of the six continental maps that are found in the atlas. The projection is announced as one of the most distinctive features of the atlas on both its front and back page. It is a very rare example of the non-academic use of minimum-error projections in contemporary atlas cartography and, as Snyder puts it, "an encouraging computer-age advance in commercial map projection choice" (Snyder, 1993, p. 246).

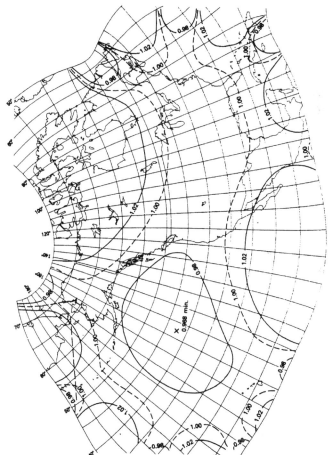

Figure 4.14 Snyder's low-error conformal map projection with irregular shaped isocols for a 50-state map of the United States (from Snyder and Voxland, 1988; reprinted with permission from Philip Voxland).

4.3   NON-CONTINUOUS   TRANSFORMATIONS

An alternative way of reducing map projection distortion consists in using different map projection equations for different parts of the Earth's surface. Some map projections of the non-continuous type have a non-interrupted graticule and are obtained by combining two pseudocylindricals along one parallel. The first to apply this type of *map projection fusion* was Goode (1925), who proposed to combine Sanson's sinusoidal projection and the Mollweide projection along the latitude of equal scale (approximately 40°44'11.98"), using the sinusoidal for the low and the Mollweide for the high latitudes. Although Goode used this so-called *homolosine* only in interrupted form (see below), the fusion produces a non-interrupted graticule with a slight break in the meridians near the 40° parallel. Erdi–Krausz (1968) took a similar approach by combining a flat-polar sinusoidal projection with the Mollweide along the parallel of 60°. Other non-interrupted projections, obtained by fusion, have been proposed by McBryde (1978), and Hatano (1972).

A popular type of non-continuous transformation consists in representing each part of the Earth's surface in relation to its own central meridian. This technique, which is known as *re-centring*, restricts the longitudinal extent within each part of the map, and so avoids extensive shearing at the outer meridians, typical of more conventional graticules. Reduction in the distortion, however, is offset by the various interruptions that result from joining of reverse-curved meridians. The use of interrupted projections being composed of a number of symmetrically arranged components goes back to the beginning of the 16th century. Yet it is only since the second half of the 19th century that non-symmetrical arrangements have been proposed with the intention of better preserving the continuity of the major continents or oceans (Dahlberg, 1962). Most of these projections have a continuous equatorial axis that connects a number of segments of either the same, or of different projections. Especially Goode's interrupted pseudocylindricals have played a key role in the promotion of such designs (Goode, 1917, 1925) (figure 4.15).

Interrupted graticules that are based on more than one type of projection, like Goode's *homolosine* (see above), are usually called *composite* map projections. Other examples of composite projections with continuity along the equator have been proposed by Watts (1970), McBryde (1978), and Baker (1986). Depending on the type of parent projections that are used also other arrangements have been suggested, including map designs with transpolar continuity (Bartholomew, 1942, 1948, 1958; William–Olsson, 1968; see also Snyder, 1993, p. 138) (figure 4.16). In some of the more recent examples that have been developed to maximally reduce distortion for each of the major land masses, different components of the projection can no longer be fitted exactly but are arranged so that they touch conveniently, either to obtain a maximum continuity of the continental area, or simply to suggest that all components ultimately fit together on the globe (see e.g. Depuydt, 1983). Sometimes the continuity between different parts of the graticule is completely renounced (Dent, 1987). For the sake of completeness it should be mentioned that some authors have also proposed interrupted map projection designs in which irrelevant areas are excluded from the map (e.g. when a map has to show

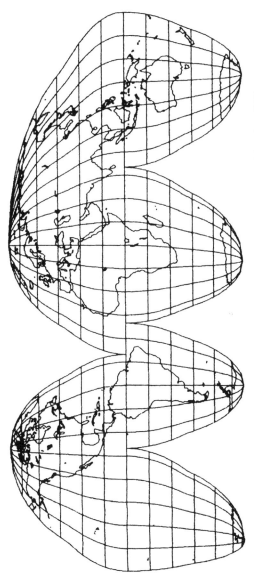

Figure 4.15 Goode's homolosine projection (from Snyder, p. 197; © 1993 by The University of Chicago. All rights reserved. Published 1993).

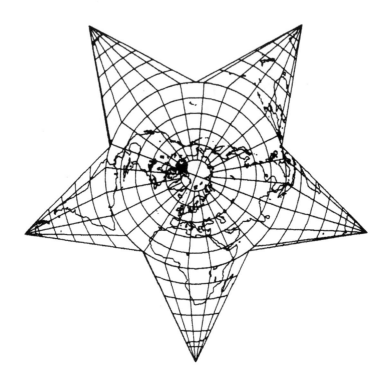

Figure 4.16 Berghaus' star projection (from Snyder, p. 139; © 1993 by The University of Chicago.
All rights reserved. Published 1993).

terrestrial distributions the oceans are left out). A well-known example of these so-
called *condensed* projections is Bomford's Oxford projection, which has been used
for world maps in Oxford atlases since 1951 (Lewis and Campbell, 1951).

Since experimentation with interrupted map projections always goes back to
the study of the distortion patterns of the non-interrupted parent graticule(s), these
projections will not be discussed further in this study. Nevertheless it should be
clear that map interruption has enormous potential. As the number of separate
components is increased, and/or the rules of continuity become less restrictive,
each part of the map can be represented with ever less distortion. Yet the question
remains if the reduction in distortion justifies the presence of discontinuities. As
will be explained in chapter 6, distortion is not the only concern in the design of a
suitable map projection, and sometimes the interruption of the Earth's surface will
not be preferred. The decision to go for an interrupted map design may be based on
the purpose of the map or the intended audience, yet much will also depend on the
personal taste of the designer. Some cartographers dislike the idea of interruptions,
saying that on interrupted graticules any suggestion of the spherical shape and
continuity of the Earth's surface is lost. On the other hand, advocates of interrupted

designs maintain that interrupted maps are a good illustration of the impossibility to represent the globe on a map without distortion.

4.4 INTRODUCING DELIBERATE DISTORTION IN MAPS

Although this monograph focuses on the reduction of distortion in maps, it should be understood that metric distortion is not the only concern in small-scale map design. As previously stated, concerns about map projection distortion will often be subordinate to general concerns about map simplicity or map readability, or to other, application-specific map requirements (see also chapter 6). Sometimes map distortion may even be used advantageously. Many examples of transformations that deliberately introduce distortion in maps exist. A distinction can be made between: (a) transformations that enlarge the central part of a map to emphasize it, or to include more cartographic detail, and (b) transformations that modify the distance between different locations, or the area of the mapping units, to make it proportional to some thematic variable. Maps which are obtained by applying the second type of transformation may look rather unusual, and are often called *cartograms*, to distinguish them from regular maps, which show, as far as possible within the limitations of scale and projection, a geometrically correct image of the Earth's surface. It should be noted, however, that the term *cartogram* is also used in a more general sense, to refer to all map-like diagrams that re-organise geographic space to emphasize or clarify a particular characteristic of a spatial process or phenomenon. Well-known examples of such cartograms are public transport maps, like Beck's famous map of the London Underground, which emphasize network connectivity, but distort actual distances and directions to increase map clarity and readability (Shirrefs, 1992; Dorling and Fairbairn, 1997, p. 44). Often the map metaphor is also used to visualise non-spatial orderings, resulting in maps of virtual spaces that are based on statistical or cognitive notions of distance or topology. Recent examples are maps of cyberspaces (Jiang and Ormeling, 2000; see also http://www.cybergeography.org/atlas/atlas.html), and interactive information maps, i.e. intuitive visual information web browsers which use the spatial concept of proximity to map large volumes of textual information according to content similarity. The closer together two items are on the map, the more similar their content (see http://www.mappamundi.net/maps/maps_009/).

4.4.1   Variable-scale maps

The most obvious way of emphasizing a selected area is to put it in the centre of the map, and enlarge it with respect to its surroundings. This also provides extra space for additional cartographic detail, which may be a good reason for applying this kind of transformation. A simple solution to accomplish the enlargement is to use an azimuthal projection and adjust its variation of scale. Azimuthal projections have a radial distortion pattern, and are completely defined by the way the radial scale varies from the centre to the boundary of the map. By choosing an appropriate function to express the radial distance from each point to the map's centre, a transformation with a suitable change of radial scale can be defined. The classic example of the use of an azimuthal projection with an enlarged centre is a map by Hägerstrand (1957) showing migration patterns near the town of Asby in central Sweden. The map was intended to portray minor movements into neighbouring towns and districts, as well as migration to distant places, and is based on an azimuthal projection in which the distance from the centre shrinks proportionally to the logarithm of the real distance. Of course, a projection of this type cannot be applied to the shortest distances. That is why on Hägerstrand's map the area within a circle of one km radius is reduced to a dot (Hägerstrand, 1957). A possible approach to construct azimuthal projections of the logarithmic type that function at any distance, is documented in Snyder (1987b).

Other examples of azimuthal projections, especially designed to emphasize a circular region, and portray it in relation to its surroundings, are Snyder's so-called *magnifying-glass azimuthal projections* (Snyder, 1987b). The idea behind these projections is to portray the central portion of the map in a regular azimuthal projection, and then gradually decrease the radial scale beyond the limits of the inner area until it reaches zero at the circular boundary of the map. Another possibility, in the case of an equal-area or equidistant map, is to represent the

Figure 4.17 "Magnifying-glass" azimuthal equal-area projection centred at 45°N, 85°W with the inner circle at 7° radius, the outer circle at 20° radius, and the area scale factor between the inner and outer circles set at 0.25 (from Snyder, 1987b, p. 64; reprinted with permission from the American Congress on Surveying and Mapping).

surrounding regions at a constant area (or radial) scale that is smaller than the scale in the central portion. This implies an abrupt change in scale at the inner boundary (figure 4.17).

Instead of transforming the geographical coordinates of spatial features directly into their corresponding map coordinates, one may also develop variable-scale maps by re-projecting the x,y-coordinates of a standard map projection in the plane. Monmonier (1977) shows several examples of non-linear reprojection of map projection coordinates, applied to a map of the 48 conterminous United States. In each example, scale is decreased from east to west to reduce the congestion of symbols in the small-sized and densely populated eastern states. The change of scale is obtained by transforming the x-coordinates of the projection, using two linear reprojection functions with different slope, joined at a hinge line (figure 4.18c), or one non-linear reprojection function with a continuously decreasing slope (figure 4.18d). The y-coordinates are uniformly scaled to fit the map frame (figure 4.18a,b). Of course, non-uniform scaling can be applied to both the x- and the y-coordinates to serve a variety of design objectives. Coordinates can also be rotated about the centre of the map before reprojection, to alter scale in any suitable direction.

Kadmon (1975) proposed another method for reprojecting the x,y-coordinates of a map, which involves a radial transformation of map projection coordinates around a chosen point so that map scale becomes inversely related to the distance from that point. The method allows exaggerated, larger-scale mapping of the area

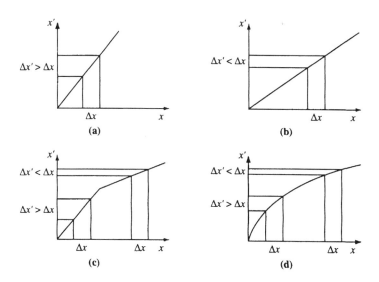

Figure 4.18 Reprojection functions with (a, b) a constant slope for a uniform increase or decrease of scale, (c) two linear slopes joined at a hinge line to yield zones of increased and decreased scale, and (d) a continuously decreasing slope for a less abrupt transition between regions of increased and decreased scale (after Monmonier, 1977).

around the focal point, and can be used for the production of town maps with crowded central areas (figure 4.19). The concept may be generalised by making the map scale proportional to a thematic variable that varies with the distance to the focal point, such as average driving time to the town centre (Kadmon, 1975). It can also be applied to produce maps with several foci, each with its own scale, proportional to some thematic variable, and its own distance decay function. In these *polyfocal maps* the decay function is considered as a potential function, which determines the influence of the focal point on its surroundings. The scale at any point on the map is obtained as the sum of the influences of all adjacent foci on that point (Kadmon and Shlomi, 1978).

Cauvin and Schneider (1989) present an alternative technique (the *piezopleth maps* method) to locally adjust the scale of a source map in relation to the spatial distribution of a thematic variable. The construction of a *piezopleth map* is accomplished in three steps. First the source map is covered with a regular mesh. Next each cell in the mesh is assigned a *thematic load*, which defines the pressure on the cell. Finally, a physical model for strength resistance calculations (the *finite*

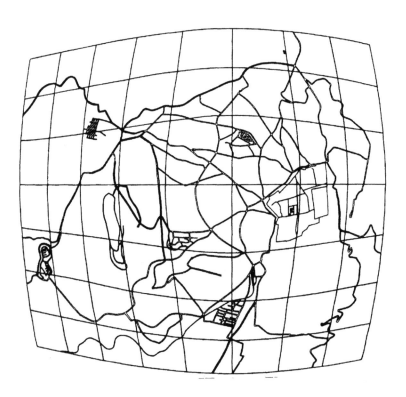

Figure 4.19 A variable-scale town map of Jerusalem with map scale inversely related to the distance from the focal point (from Kadmon, 1975, p. 51; reprinted with permission from Naftali Kadmon).

elements method) is applied, which deforms the structure of the mesh to balance the strains that are applied to each of the nodes in the network. Cells that are more heavily loaded than their neighbours will spread out, cells with a relatively small load will shrink. When applied to areal features, all cells that are part of the same feature will receive the same load. The higher the value of the thematic variable, the more an expanding area will tend to grow proportionally, from the inside out, and vice versa.

4.4.2   Distance and area cartograms

In Kadmon and Shlomi's polyfocal maps, as well as in Cauvin and Schneider's piezopleth maps, the distortion of geographic space is controlled by the spatial distribution of a thematic variable. The higher the value of the variable for a particular spatial unit, the more the area of that unit will be exaggerated on the map. However, following from the method of construction of these two types of maps, area will not be proportional to the variable. Other types of maps in which thematic variables play the leading part in determining how geographic space is portrayed are distance and area cartograms, which are based on transformation methods that seek to achieve proportionality of distances (areas) to some variable.

In *distance cartograms*, the distance between different map locations is proportional to a thematic variable that describes how close to one another two locations are in terms of travel time, travel costs, level of socio-economic interaction, etc. Two types of distance cartograms can be distinguished. The first type is used to portray the strength of the relationship between a central point and all other points on the map. The construction of this type of cartogram is often based on a generalisation of the equidistant azimuthal projection, with directions from the centre shown correctly, but with distances proportional to a thematic variable like, for example, travel time (Muller, 1978), or parcel post rates (Monmonier, 1990). If distances are defined along minimum paths in a network then the condition for azimuthality cannot be met. Clark (1977) presents a time-distance transformation of the non-azimuthal type to show shortest travel times in a transportation network from one selected node to all others.

The second type of distance cartogram takes account of the connections that exist between every pair of points in a network of locations. It may again be used to portray travel time, travel cost, actual road mileage, etc. In fact, this type of cartogram can be regarded as a special case of a map projection that minimises the distortion of distances between selected pairs of points. The construction of such a map projection involves the determination of point locations from known distances between the points. A solution, based on iterative least-squares trilateration, is found in Tobler (1977). Tobler applied the technique for the construction of maps preserving spherical or loxodromic distances (see also section 5.1), yet it may as well be used for the construction of distance cartograms. Of course, the goodness of fit between the distances shown on the cartogram, and the targeted functional distances will depend on the adequacy of Euclidean geometry as a description of the functional space. Muller (1982) shows examples of non-Euclidean distance cartograms for shortest route distances, actual travel times (by car), estimated car travel time, and air travel times between ten urban centres in Germany. Starting with a generalised version of the Minkowskian distance model, the location of the ten urban centres, and the parameters of the distance model are simultaneously adjusted until an optimal configuration is obtained (figure 4.20). The study demonstrates that the use of alternative distance measures can lead to much better approximations of geographical geometries than the use of Euclidean distance. A major problem with non-Euclidean cartograms, however, is that they are difficult to understand, and can easily be misinterpreted. Hence they are not suitable for presentation to a large public (Muller, 1982).

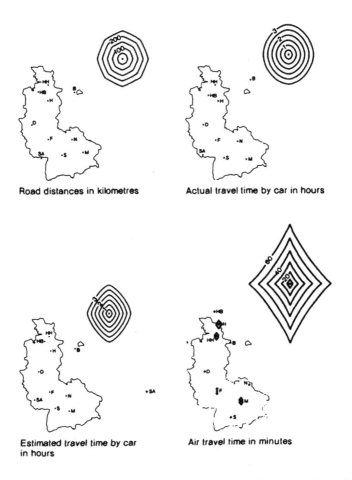

Road distances in kilometres   Actual travel time by car in hours

Estimated travel time by car   Air travel time in minutes
in hours

Figure 4.20 Non-Euclidean distance cartograms for shortest route distance, actual travel time by car, estimated travel time by car, and air travel time between ten urban centres in Germany. The scales for measuring functional distance vary according to the distance model that was used (figure 4 in "Non-Euclidean Geographic Spaces: Mapping Functional Distances" by Jean-Claude Muller, *Geographical Analysis*, Volume 14, Number 3 (July 1982) is reprinted by permission. Copyright 1982 by The Ohio State University. All rights reserved).

In *area cartograms*, the area of each mapping unit is shown proportional to some thematic variable. Area cartograms, which are sometimes also referred to as *value-by-area maps* (Raisz, 1938) or *value-by-area cartograms* (Dent, 1975), have a long history. While many variables can be used to construct them, most examples scale the map areas to be proportional to population. Using a population cartogram as a base for statistical mapping allows the map user to see the relationship between the mapped variable and the distribution of population, and eliminates the risk of underestimating the importance of small regions with high population densities. A distinction must be made between contiguous and non-contiguous area cartograms (Slocum, 1999, p. 181). Contiguous cartograms retain the contiguity of mapping units, and are in fact a special type of equal-area projection, not with a constant density of spherical surface area, like a regular equal-area projection, but with a constant thematic density distribution (Tobler, 1963). Just like on a regular equal-area projection, the shape of the units and their angular location with respect to each other will be distorted (figure 4.21). To facilitate the recognition of mapping units, the cartographer may provide cognitive map support in the form of an inset map that depicts the area on a conventional projection (Dent, 1975), or by labelling the units (Griffin, 1983). Alternatively, one may also use animation so that the user can smoothly compare the normal and transformed view of an area without having to shift gaze (Keahey, 1999). In non-contiguous cartograms, the shape of the units is retained, but gaps are introduced between the units so that contiguity is lost. To provide a suitable context for unit recognition,

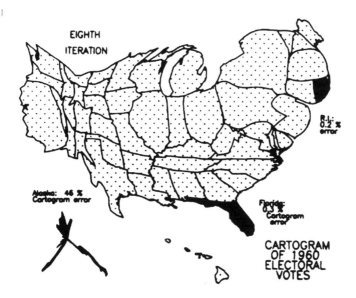

Figure 4.21 Contiguous area cartogram for the United States showing the "electoral area" for each state in the 1960 presidential election (from Dougenik *et al.*, 1985, p. 80; courtesy of Association of American Geographers).

non-contiguous cartograms can be constructed on top of a conventional map or map outline, with each unit positioned close to its original location (Olson, 1976) (figure 4.22).

Contiguous area cartograms are by far the most popular, probably because of their dramatic impact (Slocum, 1999, p. 181). Drawing these cartograms by hand is a tedious and time-consuming process, which can only be accomplished by strongly simplifying the boundaries of the individual mapping units before scaling and rearranging them. Most cartograms that are designed by hand depict mapping units (countries, states, enumeration units) as contiguous rectangles (see, for example, Kidron and Segal, 1984). Constructing such a cartogram involves frequent adjustment of the shape and the position of the rectangles, without changing their size, in an attempt to maximally preserve the topology of the original map. Apart from the impracticality of this approach, the substitution of rectangles for the original shapes of the units may hamper map unit recognition, thus reducing the communicative power of the cartogram.

In the last thirty years various computer-assisted methods have been proposed to construct contiguous area cartograms with more realistic polygon boundaries (Tobler, 1973; Selvin *et al.*, 1984; Dougenik *et al.*, 1985; Tobler, 1986; Dorling, 1995; Torguson, 1990; Gusein-Zade and Tikunov, 1993; Kocmoud, 1997; Keahey, 1999). While most methods seek to enhance the recognition of the distorted mapping units by preserving as many local features of the original boundaries as

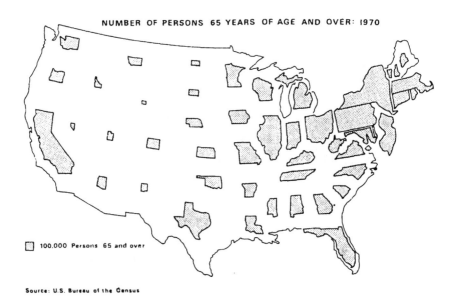

Figure 4.22 Non-contiguous area cartogram for the United States showing the number of persons 65 years of age and over for 1970 (from Olson, 1976, p. 372; courtesy of Association of American Geographers).

possible, some are more successful in attaining this goal than others. A critical review of the different approaches that have been suggested is found in Gusein-Zade and Tikunov (1993), and in Kocmoud (1997). Kocmoud identifies several shortcomings experienced by existing methods based on a list of seven criteria that should be fulfilled, and proposes a new approach for cartogram construction that satisfies each of the listed criteria. In Kocmoud's approach the automatic construction of a contiguous area cartogram is thought of as a constrained optimisation problem. First desired areas are achieved without regard to shape. Then area boundaries are adjusted in such a way that recognition is maximised, subject to the constraints that the area of each mapping unit is held at its assigned value, and that map topology is maintained. The algorithm alternates between the two goals of resizing mapping units to their correct areas, and then restoring the units' shapes while attempting to hold the areas fixed. This process is halted when the resulting map from the area-adjusting routine contains more shape error than the preceding area-adjusted map. An additional feature of Kocmoud's method is that it allows the user to adjust the level of area accuracy to obtain a better representation of shape. The user can also modify or pin down vertex locations at any time to exert control over the result. To reduce computation time, a multi-resolution mechanism is used. Initially, the optimisation method is applied to a coarsely resampled version of the original map. Once a solution is attained, the distorted map is reconstructed to its full resolution and resampled at half the current level of coarseness. Then the optimisation method is applied again. The process is repeated until a satisfactory solution with enough map detail is attained. In spite of the favourable results achieved with this method (figure 4.23), the construction of a

Figure 4.23 1996 United States population cartogram resulting from applying Kocmoud's constraint-based method (from Kocmoud, 1997, p. 65; reprinted with permission from Christopher Kocmoud).

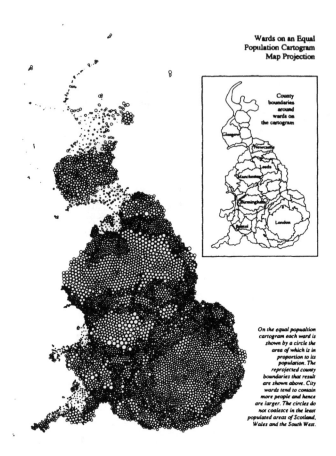

Figure 4.24 Wards on an equal population cartogram of Great Britain resulting from applying Dorling's algorithm (from Dorling, 1993, p. 171; courtesy of The British Cartographic Society).

cartogram can still take many hours, which makes the method unsuited for on-the-fly production of cartograms.

Recently, Keahey (1999) proposed an alternative method for the construction of contiguous cartograms, which repeatedly transforms an initially uniform grid until the size of each of its cells approximates a data-driven magnification value. Texture mapping is used to automatically interpolate an image of the map onto the transformed grid. The method is somewhat similar to the first algorithm for cartogram construction that was presented by Tobler in 1973. Like other methods that use some kind of rubber sheet approach (Tobler, 1973; Selvin *et al.*, 1984; Dougenik *et al.*, 1985; Gusein-Zade and Tikunov, 1993), it is prone to balloon-like distortion effects, with straight-line boundaries between mapping units warped into curves (figure 4.21). As a result, the *recognisability* of regions is not as good as on Kocmoud's cartograms. Keahey's method, however, has the distinct advantage of converging at near-interactive frame-rates, making it a useful technique for the on-line generation of cartograms, and for real-time animation.

Dorling (1993) developed an algorithm for the automatic production of a special type of non-contiguous cartogram, with circles substituted for the original shapes of the units. The algorithm begins by placing a properly scaled circle in the centre of each unit on an equal-area projection, and then gradually moves the circles away from one another so that no two overlap. Although the algorithm tries to retain a point of contact between neighbouring units, it is not always possible to meet this constraint. However, if applied to a large number of mapping units, the result may look similar to a contiguous area cartogram, with regions with a high population density expanding in a radial fashion, and less populated areas shrinking accordingly (figure 4.24). Since Dorling's cartograms typically show a large number of mapping units, and do not provide shape information, they are very difficult to interpret, even if shown in combination with a conventional base map. Interactive solutions, based on techniques for data exploration like *brushing* and dynamic linking of maps, are especially suited for examining these representations (see also http://www.geog.le.ac.uk/argus/ICA/J.Dykes/3.3.html).

CHAPTER FIVE

Development of map projections with low finite distortion

Global measures of distortion, as defined in chapters 2 and 3, are not only useful to compare the distortion on various map projections, they can also be used to optimise the parameters of a particular transformation. For instance, the aspect of a map projection can be optimised by minimising some global measure of distortion, calculated for the area of interest. Some examples of aspect optimisation, using infinitesimal measures of distortion, have been described already in chapter 2 (see section 2.6). Modifications of map projections, obtained by changing the position of the lines of zero distortion (see section 4.1.1), can be optimised in a similar way. A more challenging exercise is to optimise the parameters of a map projection transformation, subject to various constraints that are dictated by the purpose of the map (i.e. special distortion properties, geometric characteristics of the graticule).

In the previous chapter several transformation methods have been presented, which make it possible to define generalised projection systems with controllable features. Especially polynomial transformations have been shown to offer much flexibility for graticule adjustment. In this chapter, most of the transformations that have been presented will be applied to develop low-error map projections with useful properties. Attention will be focused on global as well as on continental mapping. The ultimate purpose is to propose a general strategy for the production of small-scale maps with less overall distortion, which may become part of an automated procedure for small-scale map projection selection. The basic ideas behind the selection procedure, as well as its development and implementation, will be discussed in the last chapter of this study.

5.1 MINIMUM-ERROR PROJECTIONS

Much has been published on the development of new map projections that minimise distortion within the limits of the area being mapped. In most studies, optimal values for the parameters of a map projection are obtained by minimising the sum of the squares of the errors in local scale, using one of the infinitesimal distortion measures defined in chapter 2 (see section 2.1). Projections that are derived in this manner are usually referred to as *minimum-error projections*. The majority of papers on minimum-error projections have been published between 1850 and 1950, before modern computers became available (Snyder, 1985). In these studies the problem of optimisation is necessarily limited to a small number of parameters (mostly one), and to map projections with simple mathematics.

The classical example of a minimum-error projection is Airy's azimuthal projection with a minimum "total misrepresentation" determined by "balance of errors" (see also section 2.1). Airy (1861) developed his projection by minimising the integral distortion measure defined by equation (2.13). For an azimuthal projection (2.13) simplifies to:

$$\int_{0}^{\beta} [(a-1)^2 + (b-1)^2] \sin \delta d\delta \tag{5.1}$$

where δ is the angular distance from the centre of the projection to any point on the globe, and β the angular distance from the centre to the rim of the spherical cap to be mapped. Airy had made an error in his derivations, but one year later James and Clarke (1862) presented the correct analytical solution for the radius from the projection centre (see Snyder, 1985, p. 64). In the same paper James and Clarke also applied Airy's minimum-error approach to determine the optimal location of the point of projection and the projection plane for the perspective azimuthal projection (optimal locations are listed for various values of β). Clarke's well-known *Twilight projection* with $\beta = 108°$ was presented in his contribution to the 9th edition of the *Encyclopaedia Britannica* (Clarke, 1879).

In a specialised treatise on the subject of map projection, Young (1920) proposed several minimum-error azimuthal projections, as well as a minimum-error conic projection, very similar to the equidistant conic. Young's conic projection provides the least distortion for a given pair of limiting parallels, yet it has very complicated formulas. Calculation of the radius of each parallel requires numerical integration. The development of minimum-error conformal, equal-area and equidistant conic projections is less complicated, since it only involves determination of the standard parallels. In 1916, Tsinger derived an optimal conformal conic as well as an optimal equal-area conic for Russia by minimising the overall scale error for the area. The overall scale error was obtained by dividing the area into k zones of small latitudinal extent, calculating the local scale error for the middle latitude of each zone, and averaging these values, using the area of the country within each zone as a weight factor (see equation (2.14)) (Tsinger, 1916). Almost 20 years later, Kavrayskiy (1934) applied the same technique to optimise the equidistant conic projection. For more details about the calculation method and the exact definition of the local distortion measures that were used the reader is referred to Snyder (1985, pp. 71–6).

The introduction of computers created new opportunities for research into minimum-error map projections that before then had been impossible to think of. Interesting examples are the studies by Reilly (1973), Stirling (1974), Snyder (1984a, 1986), and González-López (1995) on the development of low-error conformal map projections for irregular-shaped regions, obtained by complex-algebra transformation of standard map projections (see section 4.2.3). In these studies, optimal transformations are obtained by minimising overall distortion for a large number of points located over the entire area. Finding the coefficients of the complex-algebra polynomial that minimise distortion requires iterative solution of several simultaneous equations (see Snyder, 1985, pp. 86–90), the number of equations depending on the order of the polynomial.

Nestorov (1997) presents an alternative approach for the definition of low-error conformal map projections, which is based on the Chebyshev–Grave theorem (Chebyshev, 1856; Grave, 1896). According to this theorem the best conformal projection for mapping a specific region – i.e. the one with the least overall distortion – is one on which an isocol (line of constant scale) bounds the region. Nestorov demonstrates how the least-squares method can be used to determine the coefficients of a k-order harmonic polynomial that describes the conformal mapping of the rotational ellipsoid onto the plane, subject to the constraint that the value of the particular scale is kept constant in a number of points along the contour of the region to be mapped. Once the coefficients that satisfy the boundary condition are obtained, the complete mapping is defined, and the direct and inverse mapping equations can be determined. Nestorov developed a computer program to generate all possible variants of his *Conformal Adaptive Mapping Projection of the Rotational Ellipsoid (CAMPREL)*, which can be obtained by varying the number of contour points, the value of the particular scale along the contour, and the degree of the harmonic polynomial. For each variant the computer program determines a set of quality indices by calculating various global distortion measures. The optimal projection is selected from among thousands of projection variants by key-sorting based on the values obtained for the indices.

Attempts to develop low-error equal-area projections for arbitrarily shaped regions are less numerous, as no rigorous criteria for optimal equal-area projections have been proposed so far. One of the reasons for this is that equal-area projections are difficult to work with, and cannot benefit from the research wealth of complex analysis like conformal projections. Any theoretical study of their distortion characteristics lacks the simplicity of similar analyses of conformal projections. Of course, one might assume that on an optimal equal-area projection the region of interest is bounded by an isocol, just like on an optimal conformal projection. Based on this assumption, Snyder (1988) proposed a double transformation that converts an oblique azimuthal equal-area projection, centred on the region of interest, into an equal-area projection with oval, rectangular or rhombic isocols, depending on the value of two transformation parameters (see section 4.2.2). In a similar way, this author proposed an equal-area polynomial transformation, which makes it possible to transform an equal-area map projection into a new equal-area projection with irregular isocols (Canters, 1991) (see section 4.2.2). The transformation was applied for the development of new equal-area world maps with low finite distortion of the continental area (see section 5.3.3), and for the development of low-error equal-area projections for Eurafrica and for the European Union (Canters and De Genst, 1997) (see section 5.4.2).

Although the work on low-error equal-area projections that is discussed in this study illustrates that isocols tend to adapt themselves to the overall shape of the mapped area, there is no mathematical proof that the Chebyshev criterion holds for equal-area projections. Recent work on optimal equal-area projections by Strebe demonstrates that the optimal equal-area map of a complex region, including one or more concavities, will not, in general, be surrounded by an isocol. According to Strebe, Chebyshev's theorem only holds for convex regions, yet it can be shown that the criterion is insufficient to define the optimal equal-area projection.

Important in Strebe's work is the *torsion* metric, which he defines as the deviation of the major axis of Tissot's indicatrix from the path of the isocol. According to Strebe the absence of torsion in every point of the map is a necessary condition for an optimal equal-area projection. Strebe also explains how an optimal equal-area map is related to its conformal analogue, and generalises his results to propose that "for any optimal conformal projection, there is not only a unique, optimal area-preserving analogue, but also a continuum of aphylactic (neither conformal nor equal-area) "optimal" projections" (Strebe, pers. com., 2001). More work in this direction is definitely needed to improve our knowledge about equal-area projections, and to develop a sound theoretical basis for the development of optimal equal-area maps.

Another study that heavily relies on computing resources for the development of low-error projections is the one by Tobler (1977), in which he describes the development of an empirical minimum-error projection for the United States. Tobler obtained his projection by minimising overall error for all great-circle distances between selected points on a regular grid covering the area, using the following distortion measure:

$$E = \sum_i (s_{ij} - s'_{ij})^2 \tag{5.2}$$

with s_{ij} the spherical distance between two points on the globe, and s'_{ij} the distance between the corresponding points on the map. The resulting projection has no formulas in the usual sense, but is defined by rectangular coordinates for the points used in the analysis. To this author's knowledge, Tobler is the first who applied the principle of least squares to minimise the distortion of finite distances, using numerical techniques. Next to his projection with minimum distortion of distances he also explains how the least-squares method can be used for the development of projections with other finite distortion properties, such as minimum angular distortion in the large, or minimum distortion of distances measured along spherical loxodromes instead of along great circles (Tobler, 1977).

Some authors have also presented low-error map projections that were not obtained by applying the principle of least squares. Particularly important for this study is the work of Peters, which has been described in detail in chapter 3. As explained in section 3.4, Peters (1975, 1978) optimised the parameters for various world map projections by minimising the average distortion of a large set of finite distances, connecting 30 000 randomly chosen pairs of points. To determine optimal parameter values, Peters (1975) made use of a simplex-based algorithm for function minimisation, developed by Nelder and Mead (1965). In this chapter various new projections with low finite distortion will be presented. All projections that will be shown have been obtained by minimising the improved version of Peters' distortion measure, which was defined and subsequently applied for map projection evaluation in chapter 3 (see sections 3.5, 3.6). Optimal parameter values for each projection were determined by means of the simplex method. Before presenting the various projections, the algorithm for function minimisation will be briefly discussed, particularly in relation to its use for map projection optimisation.

5.2 MINIMISATION OF MAP PROJECTION DISTORTION USING THE SIMPLEX METHOD

Each function of n variables can be represented in a $n+1$ Cartesian coordinate system by letting each variable, as well as the function value correspond with one of the axes. If the function value is calculated for $n+1$ points in the n-dimensional factor-space, these points will form a general simplex (figure 5.1). The method for function minimisation proposed by Nelder and Mead (1965) moves an initially defined simplex through the "landscape", adapts it to the local characteristics of the function, and contracts on to the final minimum. To reach the minimum Nelder and Mead define three operations – *reflection*, *expansion* and *contraction* – which are repeatedly used in the iterative path-finding process.

If P_0, P_1, ..., P_n are the $n+1$ points in n-dimensional space forming the current simplex, then P_h and P_l are defined as the points with the highest and the lowest function value respectively. The algorithm for function minimisation starts with a reflection of P_h in the hyperplane of the remaining points. The reflection of P_h is denoted P^*. Its coordinates are defined as follows:

$$P^* = (1+\alpha)\overline{P} - \alpha P_h \qquad (5.3)$$

where $\overline{P}$ is the centroid of all P_i with $i \neq h$, and α is a positive constant, which is called the *reflection coefficient*. P^* is on the line joining P_h and $\overline{P}$, on the far side of $\overline{P}$ from P_h (figure 5.2). If the function value y^* at P^* lies between y_h and y_l, then P_h is replaced by P^* and the process is repeated with the new simplex. If $y^* < y_l$, then a new minimum has been found, and P^* is expanded to P^{**} to accelerate the minimisation process:

$$P^{**} = \gamma P^* + (1-\gamma)\overline{P} \qquad (5.4)$$

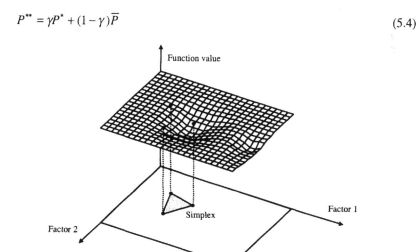

Figure 5.1 Definition of a general simplex in a two-dimensional factor-space.

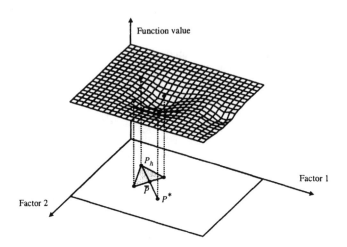

Figure 5.2 Reflection of the point with the highest function value about the centroid of the simplex.

The *expansion coefficient* γ, which is greater than unity, defines the ratio of the distance $P^{**}\overline{P}$ to $P^{*}\overline{P}$. If $y^{**} < y_l$ then P_h is replaced by P^{**} and the process is restarted. If $y^{**} \geq y_l$, then the expansion has failed, and P_h is replaced by P^{*} before restarting.

If on reflecting P to P^{*} the function value $y^{*} > y_i$ for all $i \neq h$, then P_h is first replaced by P^{*} (if $y^{*} \leq y_h$) or is left unchanged (if $y^{*} > y_h$). Then it is contracted to P^{**} in the following way:

$$P^{**} = \beta P_h + (1 - \beta)\overline{P} \tag{5.5}$$

The *contraction coefficient* β lies between 0 and 1 and is the ratio of the distance $P^{**}\overline{P}$ to $P_h\overline{P}$ (figure 5.3). P^{**} is taken as the new P_h and the process is restarted, unless $y^{**} > y_h$, which means that the contracted point is worse than P_h. In that case all vertices of the simplex are contracted in the direction of the point with the lowest function value P_l by replacing all P_i's by $(P_i + P_l) / 2$. Then the process is restarted. Figure 5.4 shows the complete flow chart of the algorithm.

Once all points are brought into the valley, the contraction operation will quickly reduce the size of the simplex, and draw all the vertices closer to the minimum. An appropriate criterion must be defined to stop the procedure. Although the criterion might be based on the size of the simplex, Nelder and Mead suggest to calculate the variance of the function values for all points of the simplex, and to halt the procedure once the variance falls below a pre-set threshold. This way the criterion is related to the curvature of the surface. If the curvature near the minimum is slight, there is no sense in finding the coordinates of the minimum very accurately. If the curvature is marked, a more precise determination of the minimum is justified. In the present study, where the function value

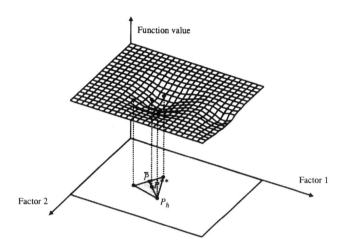

Figure 5.3 Contraction of the point with the highest function value towards the centroid of the simplex.

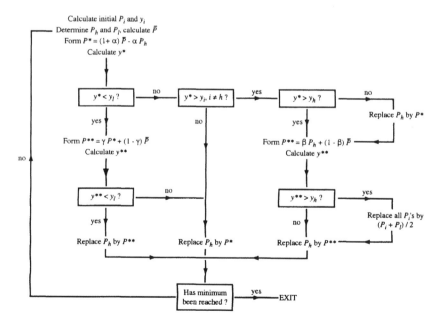

Figure 5.4 Flow chart of the simplex algorithm for function minimisation (after Nelder and Mead, 1965).

corresponds with the mean distortion of finite distances for the entire map, the threshold value for halting the procedure was set equal to 10^{-12}. The coefficients α, β, γ determine how the volume of the current simplex is changed by reflection, contraction or expansion respectively. Nelder and Mead experimented with different values for α, β, γ and found that the simple strategy $\alpha = 1$, $\beta = 0.5$ and $\gamma = 2$ was the best. These values have also been used in the present study. Other strategies proved to be more variable in performance (more frequently converging to false minima) and, on the average, slower to converge. The size, shape and orientation of the initial simplex also have an effect on the speed of convergence, and on the ability of the procedure to find the right minimum. The simplex should be adequately sized to prevent it from contracting in the wrong valley at an early stage in the process, while its range in each dimension should be adapted to the sensitivity of the function value to small changes in the value of the variable corresponding with that particular dimension. One way to build the initial simplex is by starting with a good estimate of the optimum vector P_S, taking this as the first point of the initial simplex (P_0), and defining the other P_1, ..., P_n points by adding a pre-set value r_1, ..., r_n (step-length) to the 1st, ..., nth coordinate of P_S, and leaving the other coordinates unchanged:

$$P_i = P_S + Q_i R^T \qquad\qquad (5.6)$$

with $R = (r_1, ..., r_n)$ defining the step-length for each dimension, and Q_i an $n \times n$ matrix with $q_{ii} = 1$, and all other elements equal to zero. In this study the variables for which an optimal value has to be found correspond with the parameters of the projection and/or the coefficients of the polynomial transformation that is applied. Experiments with different step-lengths, carried out in connection with the derivation of the optimised map projections discussed below, indicated that stable results are obtained with $r_i = 5°$ for the optimisation of aspect (position of meta-pole and geographical pole) and standard lines, $r_i = 0.1$ for first-order polynomial coefficients, $r_i = 0.05$ for second-order and third-order coefficients, and $r_i = 0.005$ for fourth-order and fifth-order coefficients, if the first estimate of the optimum vector P_S is well-chosen. For the optimisation of aspect and standard lines, initial values for the parameters can be estimated from the position, the shape and the orientation of the area to be mapped. For polynomial transformation, the initial value of the first-order coefficients C_{10} and C'_{01} is best set equal to one, the value of all other coefficients equal to zero. In that case the optimisation process starts with the original projection (the parent projection).

As has been shown in the previous chapter, all functional constraints that may be imposed upon a polynomial transformation can be satisfied by putting some of the coefficients equal to zero, or by defining a dependency between two or more of the coefficients. In both cases the number of dimensions of the field of search is reduced. However, as will be shown below, to derive useful graticules it will very often be necessary to define additional constraints that reduce the solution space, for example to prevent the transformed graticule from overlapping itself, to restrict the length of the pole line, and so on. In general these constraints are of the type:

$$f(C_{01}, C_{02}, ..., C_{n0}, C'_{01}, C'_{02}, ..., C'_{n0}) > 0 \qquad (5.7)$$

They can easily be introduced in the simplex procedure by modifying the function to be minimised to take a large positive value if the conditions to be met are not satisfied. Any trespassing by the simplex over the border of the solution space will then be followed by contractions that will keep it inside, provided that the initial simplex is located inside the permitted region.

The simplex method makes no assumptions about the surface except that it is continuous and one-valued. When minimising map projection distortion both conditions are satisfied. However, depending on the type of parameters to be optimised, the landscape of function values may include different local valleys. Hence one will not always be certain that the absolute minimum has been found, unless the landscape is examined in detail. If the number of parameters is limited, one may do a quick exploration of the landscape by calculating function values for fixed, not too small increments of each parameter. This will indicate where the valleys are located, and will suggest optimal locations for the definition of the initial simplex. By repeating the optimisation process for various start locations different local minima, as well as the global minimum, can be found. The explorative method may be very useful to identify alternative solutions, for example, when selecting an optimal aspect (see section 5.4.1).

For a large number of parameters, as with polynomial transformation, the systematic exploration of the landscape becomes awkward, and application of the simplex method will yield a low-error solution, which not necessarily corresponds to the minimum-error case. The examples discussed in this chapter, however, will prove that "blind use" of the simplex method allows one to derive new projections with considerably less distortion than the parent graticules that are chosen as the starting point for the optimisation. The method has the advantage that it does not require the calculation of first-order or second-order derivatives. It is computationally compact and can be applied to any map projection. Implementation is independent of the complexity of the original projection and of the number of parameters to be optimised. Constraints on the volume to be searched are easy to impose (see above). All this makes the simplex method very suited for automated selection and optimisation of map projections (see chapter 6). The only drawback is the efficiency of the method. Execution time is dependent on the complexity of the function and the total number of function evaluations. Although simplex optimisation works fast for a small number of variables, the mean number of function evaluations for convergence increases rapidly with the number of variables to be optimised. However, since for the majority of applications (see the examples below) the number of variables will be limited (usually less than 10), this is no real inconvenience.

5.3 LOW-ERROR PROJECTIONS FOR WORLD MAPS

Since the promotion of the Robinson projection (see section 3.1) by *the National Geographic Society*, as part of their campaign for geographic literacy (Garver, 1988), the awareness that a good representation of large shapes should be a key feature in the development of new world map projections has increased. In the third chapter of this study a method for the measurement of distortion in the large has been proposed that takes account of both the distortion of shape and the relative distortion of area (see sections 3.5, 3.6). This method will now be used to develop new map projections with minimum distortion of the continental masses.

First of all, several low-error pseudocylindrical and polyconic map projections will be derived, using Wagner's three transformation methods (see section 4.1.2). Results will be compared with the parameter choices proposed by Wagner himself. This will give a good indication of the flexibility of Wagner's approach, and will also demonstrate some of the benefits and limitations of parameter optimisation as applied to global mapping. Next, different types of polynomial transformation (see sections 4.1.3, 4.2) will be discussed. Since distortion on world maps is extreme, one may argue that it is of utmost importance that the minimisation of scale distortion is not hampered by geometric conditions that restrict the flexibility for graticule adjustment, and that have no practical utility. On most projections that are presently used for world maps (including Wagner's graticules), the parallels and the meridians are regular curves (straight lines, circular arcs, semi-ellipses or well-defined portions of ellipses, sinusoids, ...). None of these conditions (except for the straightness of the parallels) proves interesting from a practical point of view, yet they all imply a lowering of the degree of freedom for optimisation of the graticule. Polynomial transformations offers maximum flexibility for adjustment of the graticule, yet it remains to be seen if optimisation of the polynomial coefficients, using the distortion measure presented in chapter 3, will also produce better world maps. In discussing the use of polynomial transformations, first attention will be paid to the development of projections with no special distortion properties, which is the type of projection that is often preferred for a world map, for its balancing of angular and area distortion (see section 2.4.2). Then the special case of equal-area projections will be considered.

5.3.1 Low-error projections obtained by applying Wagner's transformation methods

Wagner proposed three transformation methods, which he applied to different parent projections to obtain a total of nine graticules that are all named after him (see section 4.1.2). Parameter values for these graticules were obtained by specifying simple geometric conditions, including the ratio of the axes, the relative length of the pole line, the scale factor along the equator, the curvature of the parallels, and the distortion of area along the parallel of 60°. However, instead of obtaining the parameter values by constraining the geometry of the graticule, one may also choose to determine parameter values through optimisation of some globally defined distortion measure.

We will now discuss optimised versions for all six transformations defined in section 4.1.2. Each of the graticules that will be shown has been obtained by minimisation of the improved version of Peters' finite distortion measure (see section 3.5), calculated for a set of 5000 distances covering the whole continental area. The algorithm for selecting the set of distances is described in detail in section 3.6. Solutions have been obtained by applying the simplex method (see section

5.2). Initial experiments with parameter optimisation indicated that Antarctica is situated so eccentrically that it severely worsens the representation of the other continents, if included in the analysis. Therefore, it was not considered in the optimisation. This explains its highly distorted appearance in some of the graticules that are presented below.

Pseudocylindrical graticules

Figure 5.5 shows the three pseudocylindrical map projections by Wagner that have already been discussed in chapter 4, together with an alternative version of each projection, obtained by optimisation of the parameter values in the general transformation formulas (see section 4.1.2). Table 5.1 lists the mean distortion of finite distances for the continental area, the parameter values (m_1, m_2, n, k_1, k_2), and the characteristics of the graticule (c = length of the pole line : length of the equator, p = length of the central meridian : length of the equator) for all six projections. For the first two transformations the parameter m_1 in table 5.1 corresponds to the parameter m in the formulas. For the generalised equal-area transformation (Wagner II) the parameter m_2, which determines the amount of area distortion, was left unchanged in order to obtain a graticule with the same area distortion as the original projection. Since all three transformation, methods are over-defined, the value of k_2 in the first transformation and the value of k_1 in the second and third transformation were set equal to one, while the values of the other parameters were optimised (see section 4.1.2).

What is most striking about the optimised projections is the length of the pole line, and the slight curvature of the meridians. All three projections have a pole line that is about 0.7 times the length of the equator (instead of 0.5 times for Wagner's choice of parameter values). Increasing the relative length of the pole line is accomplished by a decrease of the value of m_1, which implies that the whole world is mapped onto a sub-zone of the parent graticule (Apianus, sinusoidal projection) of smaller latitudinal extent (see section 4.2.1). While this automatically reduces the curvature of the meridians in the higher latitudes, an even more substantial reduction of the curvature of the meridians is obtained by also lowering the value of the parameter n. This limits the longitudinal extent of the section of the parent projection onto which the entire world is mapped. As can be seen from table 5.1, all three optimised graticules have lower values for m_1 and n than the original projections that have been proposed by Wagner.

The increased length of the pole line, and the moderate curvature of the meridians makes that the optimised graticules show a less marked increase in angular distortion from the centre towards the NW, NE, SW and SE corners of the graticule. A comparison of the patterns of angular distortion for the optimised graticules (figure 5.6b, figure 5.7, figure 5.8b) with those obtained for Wagner's original projections (figure 4.2b, figure 4.3, figure 4.4b) illustrates this. The shape of North America and Australia is better retained in the optimised projections than in the original ones.

For the optimised versions of Wagner I and Wagner II, which both have fixed area distortion (zero for the entire graticule for Wagner I, and increasing from zero at the equator to a value of 1.2 at the parallel of 60° for Wagner II), the ratio of the

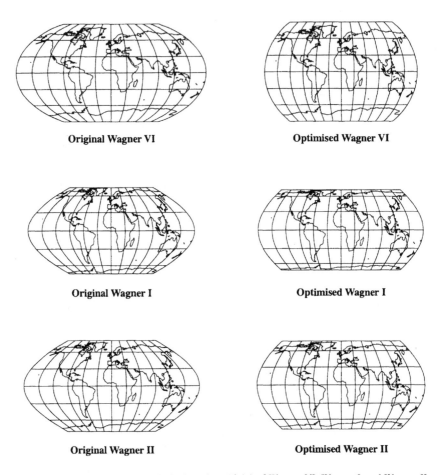

Original Wagner VI   Optimised Wagner VI

Original Wagner I   Optimised Wagner I

Original Wagner II   Optimised Wagner II

Figure 5.5 Original (left) and optimised versions (right) of Wagner VI, Wagner I, and Wagner II
(all graticules are on the same scale).

axes is almost correct. For the optimised Wagner VI, however, the central meridian
is 0.64 times the length of the equator. This results in a graticule with the lowest
area and angular distortion in the middle latitudes, where most of the continental
area is located. As has already been shown in chapter 3, all standard map
projections that are highly ranked in terms of overall distortion have similar
distortion patterns, and show the same E–W compression in the lower latitudes
(see section 3.5). Compared to these projections, the optimised Wagner VI has a
mean distortion value that comes very close to the minimum value obtained for the
Winkel–Tripel (1.1420).

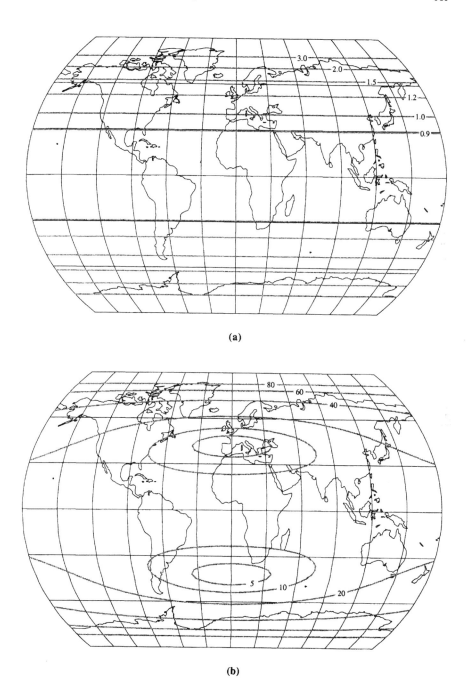

(a)

(b)

Figure 5.6 Optimised version of Wagner VI with lines of constant area scale (a), and lines of constant maximum angular distortion (b).

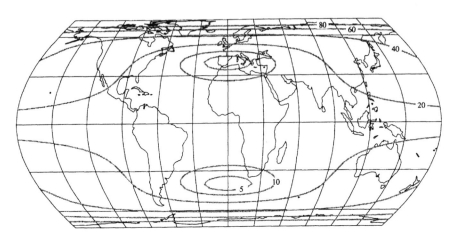

Figure 5.7 Optimised version of Wagner I with lines of constant maximum angular distortion.

Polyconic graticules

Figure 5.9 shows Wagner's three polyconic projections. On the left are Wagner's original projections, on the right the projections that are obtained by parameter optimisation. Table 5.2 lists the mean finite scale factor, the parameter values and the ratio of the axes (p) for all six projections. For the first two transformations the parameter m_1 again corresponds to the parameter m in the transformation formulas. Also this time the value of the parameter m_2 in Wagner's third transformation method has been left unchanged to obtain a graticule with the same area distortion pattern as the original Wagner VIII projection.

The most striking about the optimised graticules is the increased curvature of the parallels, which is obtained by increasing the value of the parameter n by a factor of about two, and this without a comparable decrease in the value of the parameter m_1. While for the construction of Wagner's original graticules the entire globe is mapped on a longitudinal strip of the transverse azimuthal equidistant or equal-area projection, extending about 60° east and west of the central meridian, for the optimised graticules the width of the strip is raised to 87° degrees on each side of the central meridian for the optimised Aitoff–Wagner, and to 126° and 119° respectively for the optimised versions of Hammer–Wagner and Wagner VIII. As a result, the pole line for the latter two projections is strongly curved and even bends back in the direction of the central meridian. The reason for this strong curvature is obviously to reduce the amount of angular distortion in the NE and NW corners of the graticule. While for pseudocylindrical projections an increase in the length of the pole line is the only available option to reduce angular distortion in the areas that are close to the outer meridian, for polyconic projections the adjustment of the curvature of the parallels offers a second means of controlling the distortion characteristics of the projection in the periphery.

A comparison of the distortion patterns for the three optimised graticules

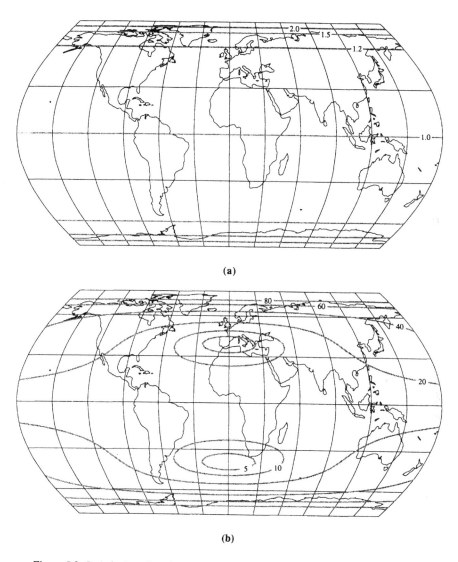

(a)

(b)

Figure 5.8 Optimised version of Wagner II with lines of constant area scale (a), and lines of
constant maximum angular distortion (b).

(figure 5.10, figure 5.11, figure 5.12) with those obtained for Wagner's original
projections (figure 4.6, figure 4.8, figure 4.9) shows a considerable improvement
in the distribution of angular distortion. For the Aitoff–Wagner projection, the
marked increase in angular distortion away from the central meridian, which is
found on Wagner's graticule (figure 4.6b), has disappeared on the optimised

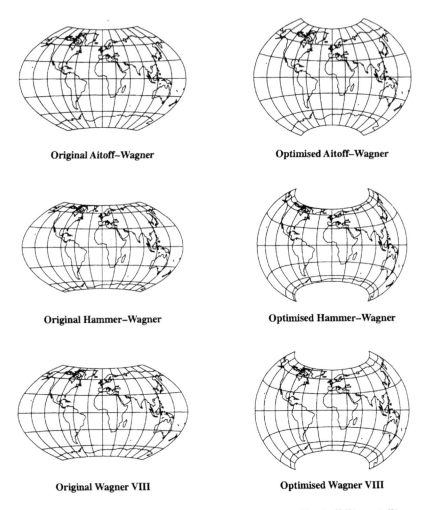

Original Aitoff–Wagner   Optimised Aitoff–Wagner

Original Hammer–Wagner   Optimised Hammer–Wagner

Original Wagner VIII   Optimised Wagner VIII

Figure 5.9 Original (left) and optimised versions (right) of Wagner IX (Aitoff–Wagner), Wagner VII (Hammer–Wagner), and Wagner VIII (all graticules are on the same scale).

graticule, which has a more rectangular pattern of distortion (figure 5.10b). Maximum angular distortion in the NE, NW, SE and SW corners of the graticule is well below 40° at the cost of a slight increase of area distortion. A rectangular-like pattern of distortion is also obtained for the optimised versions of Hammer–Wagner (figure 5.11) and Wagner VIII (figure 5.12b). The distortion patterns for these two projections in particular show very well how the optimisation of projection parameters is controlled by the spatial distribution of the continents. The presence of five foci of low distortion, one in the centre and

Table 5.1 Mean finite scale factor, parameter values, and graticule characteristics for original and optimised versions of Wagner's pseudocylindrical projections.

	Mean finite scale factor	m_1	m_2	n	k_1	k_2	c	p
Wagner VI	1.177	0.8660		0.8660	1.0000	1.0000	0.5000	0.5000
Wagner I	1.179	0.8660		0.6667	1.0000		0.5000	0.5000
Wagner II	1.165	0.8802	0.8855	0.6667	1.0000		0.5000	0.5000
Optimised Wagner VI	1.146	0.7415		0.7198	0.8089	1.0000	0.6710	0.6368
Optimised Wagner I	1.169	0.6938		0.5158	1.0000		0.7202	0.4732
Optimised Wagner II	1.161	0.7223	0.8855	0.4921	1.0000		0.7035	0.5113

Table 5.2 Mean finite scale factor, parameter values, and graticule characteristics for original and optimised versions of Wagner's generalised polyconic projections.

	Mean finite scale factor	m_1	m_2	n	k_1	k_2	p
Aitoff–Wagner	1.149	0.7778		0.2778	1.4725	1.6733	0.5682
Hammer–Wagner	1.164	0.9063		0.3333	1.4660		0.5000
Wagner VIII	1.149	0.9212	0.8855	0.3333	1.4660		0.5000
Optimised Aitoff–Wagner	1.135	0.7459		0.4809	1.0226	1.2664	0.5989
Optimised Hammer–Wagner	1.121	0.8034		0.6977	1.0363		0.4709
Optimised Wagner VIII	1.119	0.8370	0.8855	0.6603	1.0304		0.5088

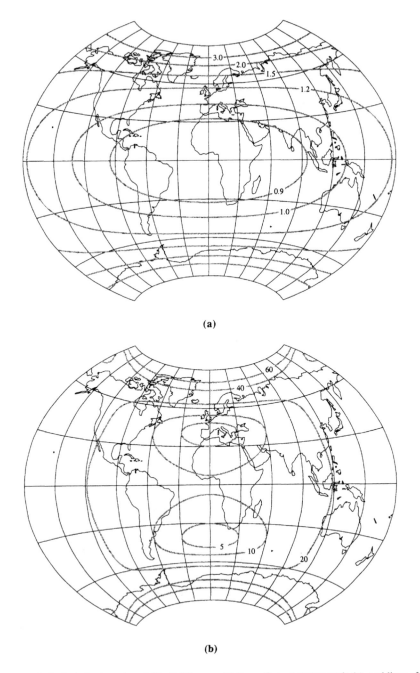

(a)

(b)

Figure 5.10 Optimised version of Aitoff–Wagner with lines of constant area scale (a), and lines of constant maximum angular distortion (b).

four in thecorners of the graticule, makes that angular distortion is less than on any standard equal-area projection or any standard projection with low area distortion.

The general improvement in the distribution of distortion is also reflected in the mean finite scale factor for the three optimised projections (table 5.2). For all three projections, the mean finite scale factor is below the value obtained for the Winkel–Tripel projection, which has been ranked highest of all standard map projections (table 3.5). What is also remarkable is the fact that the mean finite scale factors for the optimised Hammer–Wagner and for the optimised Wagner VIII, are lower than for the optimised Aitoff–Wagner, which may lead to the conclusion that the first two projections are to be preferred above the third one. A quick look at the distortion patterns, however, shows that for Hammer–Wagner and Wagner VIII the overall reduction of distortion is achieved at the cost of higher distortion values in the polar regions, and in areas with a low concentration of land, while for the optimised Aitoff–Wagner a much more balanced pattern of distortion is obtained. This confirms some of the earlier remarks about the distribution of distortion on equal-area projections, as compared to projections with intermediate distortion characteristics, and again shows that the selection of a map projection cannot be based solely on the value of one global measure of distortion.

5.3.2 Low-error polynomial projections with intermediate distortion characteristics

As has been explained in detail in chapter 4, a whole family of new map

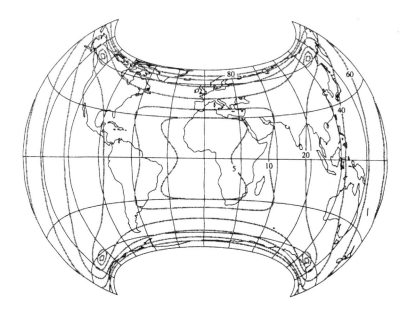

Figure 5.11 Optimised version of Hammer–Wagner with lines of constant maximum angular distortion.

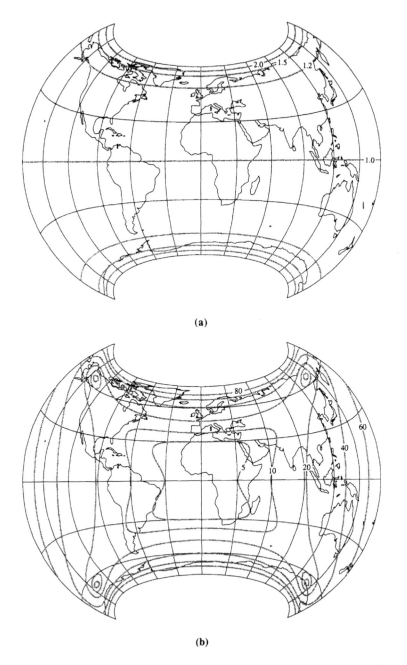

(a)

(b)

Figure 5.12 Optimised version of Wagner VIII with lines of constant area scale (a), and lines of constant maximum angular distortion (b).

projections with controllable geometry can be defined by expressing the relationship between the geographical latitude and longitude and the coordinates in the map plane by means of power series (see section 4.1.3). In its most general form, i.e. if no special distortion properties are required, the map projection process can be described by the following polynomial transformation:

$$x = Rf_1(\lambda, \phi)$$
$$y = Rf_2(\lambda, \phi)$$

(5.8)

with

$$f_1(\lambda, \phi) = \sum_{i=0}^{n} \sum_{j=0}^{n-i} C_{ij} \lambda^i \phi^j$$

(5.9)

and

$$f_2(\lambda, \phi) = \sum_{i=0}^{n} \sum_{j=0}^{n-i} C'_{ij} \lambda^i \phi^j$$

(5.10)

with x,y the map projection coordinates, λ and ϕ the geographical longitude and latitude, R the radius of the generating globe, and C_{ij} and C'_{ij} the polynomial coefficients defining the properties of the graticule. As has been shown, most geometric conditions that may be useful in cartographic practice (level of symmetry, spacing of the parallels and the meridians, length of the pole line, ratio of the axes, ...) can be imposed by putting proper restrictions on the values of the polynomial coefficients. For the derivation of new world maps with low distortion, the use of fifth-order polynomials ($n = 5$) seems appropriate. As has been explained before, fifth-order polynomials can be utilised to approximate any arbitrary world map projection to a degree of accuracy that is more than adequate for map reproduction (see section 3.1). As such these polynomials can be expected to have sufficient flexibility for the development of new graticules that satisfy any appropriate combination of constraints.

Low-error polyconic projections with pole line

In the following, several low-error projections with curved parallels and meridians will be presented. Starting from a maximum level of flexibility (non-constrained optimisation), the number of polynomial coefficients to be optimised will be systematically reduced by the introduction of new constraints. This will take us from highly irregular graticules to more conventional ones. Again optimisation will be based on the minimisation of Peters' improved distortion measure (see sections 3.4, 3.5), using a set of 5000 distances randomly distributed over the continental surface, and applying the simplex method. The coefficient values corresponding to the *Plate Carrée* (C_{10} and C'_{01} equal to one, all other coefficients equal to zero) will define the first point of the initial simplex (see section 5.2).

Antarctica will not be considered in the optimisation for the same reasons as before, unless indicated otherwise.

Optimisation without constraints

If no particular constraints are imposed then optimisation of (5.9) and (5.10) involves no less than 40 (!) independent coefficients (C_{00} and C'_{00} becoming zero if the origin of the coordinate system is chosen to coincide with the intersection of the Greenwich meridian and the equator). Figure 5.13 shows the obtained graticule. Table 5.3 lists the values of the polynomial coefficients. Since only the position of the origin has been fixed, there is no doubt that the graticule in figure 5.13 will have the lowest overall distortion of all normal aspects with central meridian at $0°$ that can be derived from (5.9) and (5.10). Due to the irregular distribution of the landmasses the equator deflects to the north, which improves the representation of the Northern Hemisphere. The southern continents, however, suffer from severe stretching in the E–W direction. Equally disturbing are the overlaps which occur outside the continental area, and which would not be permissible on a finished map. To avoid these overlaps, it usually suffices to impose simple constraints on the form of the pole line. All overlaps that initially occurred for some of the graticules described below have been avoided on the final maps by imposing $\partial x/\partial \lambda$

Table 5.3 Coefficient values and mean finite scale factor for the non-constrained low-error polyconic projection.

$C_{10} = 1.0542$	$C'_{10} = -0.0009$
$C_{01} = -0.0173$	$C'_{01} = 0.9744$
$C_{20} = 0.0088$	$C'_{20} = 0.2607$
$C_{11} = -0.4146$	$C'_{11} = -0.0040$
$C_{02} = 0.0036$	$C'_{02} = 0.0712$
$C_{30} = -0.0449$	$C'_{30} = 0.0005$
$C_{21} = 0.0010$	$C'_{21} = -0.0708$
$C_{12} = -0.0432$	$C'_{12} = -0.0002$
$C_{03} = 0.0055$	$C'_{03} = -0.0694$
$C_{40} = 0.0001$	$C'_{40} = -0.0081$
$C_{31} = 0.0144$	$C'_{31} = 0.0001$
$C_{22} = -0.0007$	$C'_{22} = -0.0146$
$C_{13} = -0.0499$	$C'_{13} = 0.0004$
$C_{04} = 0.0053$	$C'_{04} = -0.0207$
$C_{50} = 0.0007$	$C'_{50} = -0.0001$
$C_{41} = -0.0006$	$C'_{41} = 0.0047$
$C_{32} = -0.0042$	$C'_{32} = 0.0003$
$C_{23} = -0.0020$	$C'_{23} = -0.0298$
$C_{14} = 0.0388$	$C'_{14} = 0.0014$
$C_{05} = 0.0017$	$C'_{05} = 0.0491$

$K_1 = 1.0465$

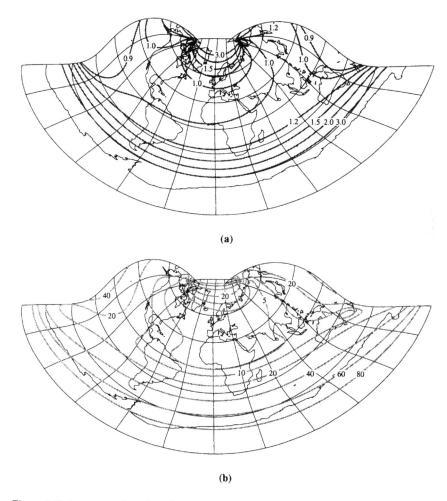

(a)

(b)

Figure 5.13 Low-error polyconic projection obtained through non-constrained optimisation with lines of constant area scale (a), and lines of constant maximum angular distortion (b).

≥ 0 and $\partial y/\partial \lambda \geq 0$ for ($\lambda = \pi$, $\phi = \pi/2$), as well as $\partial x/\partial \lambda \geq 0$ and $\partial y/\partial \lambda \leq 0$ for ($\lambda = \pi$, $\phi = - \pi/2$) for graticules that are not symmetrical about the equator.

Introduction of geometric constraints

The low-error map shown in figure 5.13 is almost symmetrical about the zero meridian. One may therefore expect the graticule not to change much, and the overall distortion value not to be much higher if symmetry about the y-axis is imposed. Doing so, however, will substantially reduce the number of coefficients to be optimised (see section 4.1.3). The strong hemispherical bias that is present in the graticule, and that is due to the uneven distribution of the continental masses, can be partly removed by forcing the equator to coincide with the x-axis. This again will lead to a reduction in the number of coefficients. Imposing both constraints, and letting the equator be equally divided by the meridians as on the globe, equations (5.9) and (5.10) respectively simplify to:

$$f_1(\lambda,\phi) = C_{10}\lambda + C_{11}\lambda\phi + C_{12}\lambda\phi^2 + C_{31}\lambda^3\phi + C_{13}\lambda\phi^3 + C_{32}\lambda^3\phi^2 + C_{14}\lambda\phi^4 \quad (5.11)$$

$$f_2(\lambda,\phi) = C'_{01}\phi + C'_{02}\phi^2 + C'_{21}\lambda^2\phi + C'_{03}\phi^3 + C'_{22}\lambda^2\phi^2 + C'_{04}\phi^4$$
$$+ C'_{41}\lambda^4\phi + C'_{23}\lambda^2\phi^3 + C'_{05}\phi^5 \quad (5.12)$$

The optimised graticule is shown in figure 5.14. Table 5.4 summarises the values for the 16 coefficients that define the graticule. As can be seen the representation of the Northern and Southern Hemisphere is much more balanced now. The non-symmetrical shape of the graticule clearly reflects the strong differences in the distribution of the landmasses in both hemispheres.

Except for the cordiform maps of the 16th century (Kish, 1965), world map projections that lack symmetry about the equator, like the ones presented above, have been proposed only occasionally. The best known examples are probably the *loximuthal* projection, which represents all loxodromes through a chosen centre as straight lines of true length (Siemon, 1935; Tobler, 1966a), Raisz' *orthoapsidal* double projections, including the famous *Armadillo* projection (Raisz, 1943; see also section 1.4.1), and Hill's *eucyclic* projection (Snyder, 1994). Obviously, for general-purpose world maps the lack of symmetry about the equator is not appreciated. Almost all projections that are used for world maps, and that represent the world without interruption, are symmetrical about both the equator and the central meridian. Adding this to the former constraints, (5.11) and (5.12) respectively become:

Table 5.4 Coefficient values and mean finite scale factor for the low-error polyconic projection with straight equator and symmetry about the central meridian.

$C_{10} = 0.8735$	$C'_{01} = 1.0065$
$C_{11} = -0.0793$	$C'_{02} = 0.0236$
$C_{12} = -0.1630$	$C'_{21} = 0.0483$
$C_{31} = -0.0111$	$C'_{03} = -0.0583$
$C_{13} = 0.0958$	$C'_{22} = -0.0009$
$C_{32} = -0.0030$	$C'_{04} = 0.0159$
$C_{14} = 0.0034$	$C'_{41} = -0.0004$
	$C'_{23} = -0.0104$
	$C'_{05} = -0.0074$

$$K_1 = 1.1083$$

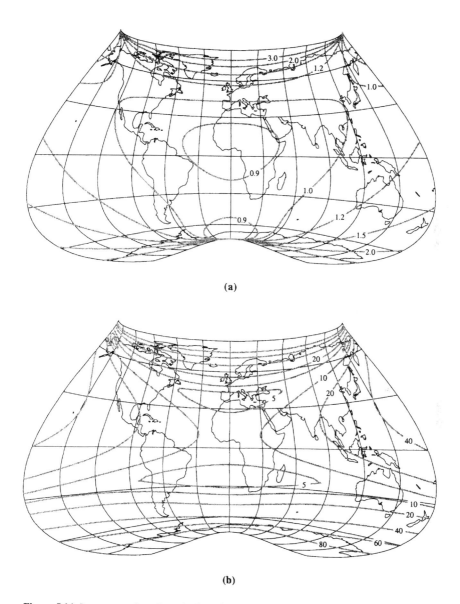

(a)

(b)

Figure 5.14 Low-error polyconic projection with straight equator and symmetry about the central meridian, with lines of constant area scale (a), and lines of constant maximum angular distortion (b).

$$f_1(\lambda, \phi) = C_{10}\lambda + C_{12}\lambda\phi^2 + C_{32}\lambda^3\phi^2 + C_{14}\lambda\phi^4 \tag{5.13}$$

$$f_2(\lambda, \phi) = C'_{01}\phi + C'_{21}\lambda^2\phi + C'_{03}\phi^3 + C'_{41}\lambda^4\phi + C'_{23}\lambda^2\phi^3 + C'_{05}\phi^5 \tag{5.14}$$

Optimisation of the 10 coefficients (table 5.5) yields the map shown in figure 5.15. As a direct result of the symmetry about the equator, the distribution of the landmasses in the Northern Hemisphere has a strong impact on the appearance of the whole graticule. The projection has a long pole line, which allows the meridians and the parallels to intersect at almost straight angles. To avoid extreme E–W stretching in the polar areas the ratio of the axes is adjusted (equator < 2× central meridian). Area exaggeration in the high latitudes is kept within acceptable bounds by a decrease in the spacing of the parallels towards the poles. The scale variation along the central meridian leads to a substantial distortion of shape in the higher latitudes (E–W stretching) and close to the equator (N–S stretching). Distortion properties are best in the middle latitudes, where most of the landmasses occur.

To obtain graticules with less variation in scale error additional constraints will have to be imposed. Equal spacing of the parallels along the central meridian seems the most obvious choice. As has been said in chapter 2, graticules with equally spaced parallels are very popular for world maps. They prove to offer a good balance between the distortion of angles and areas, and have a pleasing effect on the overall representation of the continents due to the absence of scale variation along the central meridian (see section 2.5.1). An equal spacing of the central meridian is obtained by letting the y-coordinate become a linear function of the latitude for $\lambda = 0$. Hence equation (5.14) simplifies to:

$$f_2(\lambda, \phi) = C'_{01} \phi + C'_{21} \lambda^2 \phi + C'_{41} \lambda^4 \phi + C'_{23} \lambda^2 \phi^3 \qquad (5.15)$$

Optimisation of the eight remaining coefficients (table 5.6) produces the graticule shown in figure 5.16. It has a mean distortion value that is lower than for standard projections with curved parallels and meridians (table 3.5). The graticule strongly resembles the Winkel–Tripel projection, except for the curvature of the pole line. Both projections have slightly curved parallels, a pole line shorter than half the length of the equator, and a ratio of the axes (equator : central meridian) less than 2:1 (1.7017 for the optimised graticule, 1.6365 for the Winkel–Tripel). The shortening of the pole line, compared to the previous graticule (figure 5.15), is a logical consequence of the equal spacing of the central meridian. Since the scale

Table 5.5 Coefficient values and mean finite scale factor for the low-error polyconic projection with twofold symmetry.

$C_{10} = 0.8202$	$C'_{01} = 1.0101$
$C_{12} = -0.1295$	$C'_{21} = 0.0398$
$C_{32} = -0.0091$	$C'_{03} = -0.0165$
$C_{14} = 0.0358$	$C'_{41} = 0.0001$
	$C'_{23} = -0.0118$
	$C'_{05} = -0.0071$

$K_1 = 1.1178$

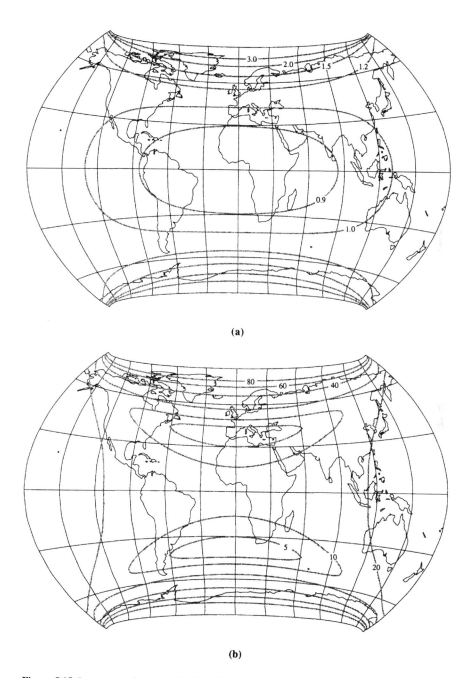

(a)

(b)

Figure 5.15 Low-error polyconic projection with twofold symmetry, with lines of constant area scale (a), and lines of constant maximum angular distortion (b).

Table 5.6 Coefficient values and mean finite scale factor for the low-error polyconic projection with twofold symmetry and equally spaced parallels.

$C_{10} = 0.8478$	$C'_{01} = 0.9964$
$C_{12} = -0.1702$	$C'_{21} = 0.0313$
$C_{32} = -0.0062$	$C'_{41} = 0.0009$
$C_{14} = 0.0036$	$C'_{23} = -0.0100$

$K_1 = 1.1315$

along the y-axis is kept constant, the shortening of the pole line is the only way that is left to reduce the exaggeration of area in the higher latitudes. To limit shape distortion near the edges of the map, the curvature of the pole line is increased. These observations show very well how all characteristics of the graticule are interrelated. Changing one property of the graticule (e.g. spacing of the parallels, length of the pole line, ...) will automatically have an impact on other features if distortion is to be maximally reduced.

Finally, the length of the axes of the projection can be adjusted to obtain an equator and a central meridian that are in a ratio of 2:1, as on the globe. Although the ratio of the axes on figure 5.16 ensures minimum overall distortion of the continental area, it compresses the equatorial zone in the E–W direction, which results in an unpleasant distortion of shape for Africa, South America, and Australia. Due to the central position of these continents on conventional maps, comparatively small distortions of these regions are well perceived by the map user. By imposing a 2:1 ratio of the axes, the shape of the landmasses in the lower latitudes can be improved at the cost of a more substantial E–W stretching of the polar areas. The adjustment can be accomplished by putting C_{10} in (5.13) equal to C'_{01} in (5.15). The optimised graticule (figure 5.17, table 5.7) has a higher mean finite scale distortion than the previous one. However, the correct ratio of the axes has a pleasing effect on the representation of the equatorial areas. Which of both graticules is to be preferred will depend on the personal judgement of the map maker.

All optimised polyconic projections that have just been described were first presented in 1989 (Canters, 1989). Since then the last graticule (figure 5.17) has repeatedly been used for world mapping by Belgian publishers. It was adopted for the first time in 1991 for the development of a new physical world map, intended for use in the first three years of secondary education, by *the Department of Development Co-operation* of the *Ministry of Foreign Affairs of Belgium (ABOS)*. Since then the graticule is known as the *Canters projection*, and has been introduced in several Belgian school atlases and geography textbooks. In the beginning of 1997, the projection was used for a new world map, showing the Human Development Index for all countries, published by the *Belgian National Centre for Development Co-operation (NCOS)*. The event attracted considerable attention in the Belgian press, and introduced the projection to a broad audience.

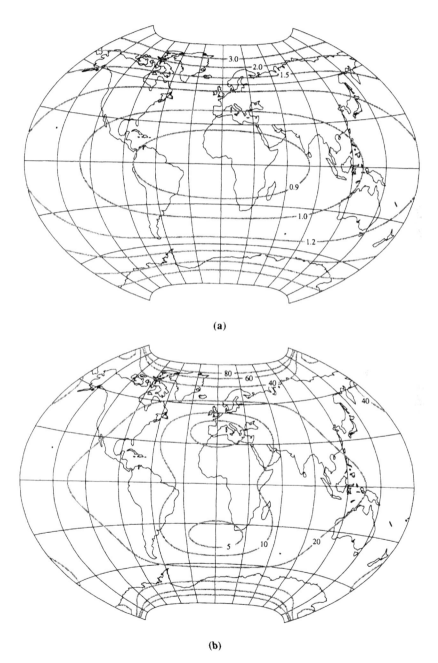

(a)

(b)

Figure 5.16 Low-error polyconic projection with twofold symmetry and equally spaced parallels, with lines of constant area scale (a), and lines of constant maximum angular distortion (b).

Low-error pseudocylindrical projections with pole line

All optimised graticules that have been described so far have curved parallels and curved meridians. However, in chapter 4 it has been shown that polynomial transformation can also be applied to develop other types of projections, if the transformation equations are properly adjusted, and the right constraints are imposed. This will be demonstrated for the pseudocylindrical class of projections, which is very popular for the mapping of world distributions. As explained in section 1.4.1, the normal aspect of a true pseudocylindrical projection has straight horizontal parallels of different length that are equally divided by the curved meridians. Hence the *y*-coordinate is a function of the latitude only, and the *x*-coordinate is a linear function of the longitude. Since the *Plate Carrée* satisfies both these conditions, pseudocylindrical projections can be derived from equations (5.9) and (5.10), by making all terms including higher powers of the longitude in (5.9), and all terms including the longitude in (5.10), equal to zero. Adding symmetry about the two coordinate axes to the base conditions, (5.9) and (5.10) simplify to:

$$f_1(\lambda,\phi) = \lambda(C_{10} + C_{12}\phi^2 + C_{14}\phi^4) \tag{5.16}$$

$$f_2(\phi) = C'_{01}\phi + C'_{03}\phi^3 + C'_{05}\phi^5 \tag{5.17}$$

Optimisation of the six coefficients (table 5.8) yields the graticule shown in figure 5.18. The projection has a mean finite scale distortion that is lower than for all pseudocylindrical projections that have been discussed and evaluated in chapter 3 (table 3.5). Just like the optimised polyconic projection with non-equally-spaced parallels (figure 5.15), the graticule has a long pole line, which indicates that the minimisation process attempts to eliminate the presence of sharp intersections of meridians and parallels throughout the map area. The central meridian of the graticule is again more than half the length of the equator, which leads to a relative N–S stretching of the equatorial areas. The spacing of the parallels decreases towards the poles to avoid extreme area distortion in the higher latitudes.

A better representation of the higher latitudes may be obtained by a shortening of the pole line. Many of the pseudocylindrical projections that have been proposed in the past have a pole line that is half the length of the equator, and have been suggested as an intermediate solution between the cylindrical projection

Table 5.7 Coefficient values and mean finite scale factor for the low-error polyconic projection with twofold symmetry, equally spaced parallels and a correct ratio of the axes.

$C_{10} = 0.9305$	$C'_{01} = 0.9305$
$C_{12} = -0.1968$	$C'_{21} = 0.0394$
$C_{32} = -0.0067$	$C'_{41} = 0.0005$
$C_{14} = 0.0076$	$C'_{23} = -0.0115$
$K_1 = 1.1454$	

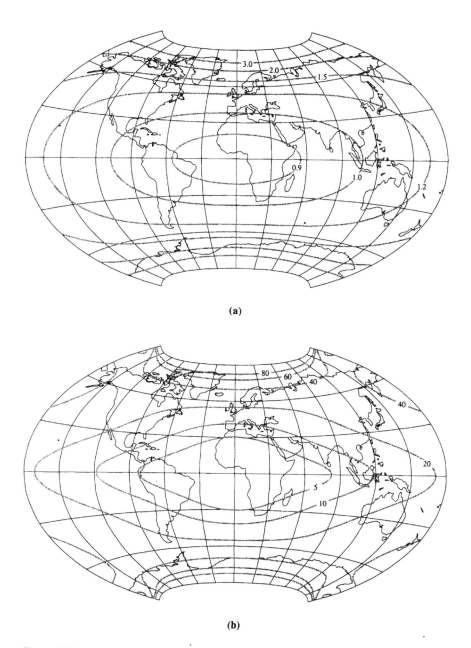

(a)

(b)

Figure 5.17 Low-error polyconic projection with twofold symmetry, equally spaced parallels, and a correct ratio of the axes, with lines of constant area scale (a), and lines of constant maximum angular distortion (b).

with its excessive E–W stretching, and the pointed-polar projection with its high compression of the polar areas (see chapter 2). To derive an optimised alternative for these projections the polynomial transformation defined by equations (5.16) and (5.17) can be applied, yet on the condition that:

$$C_{14} = \left[-C_{10} - 2\left(\frac{\pi}{2}\right)^2 C_{12} \right] \Big/ 2\left(\frac{\pi}{2}\right)^4 \qquad (5.18)$$

As before, a good starting point for optimisation is obtained by putting C_{10} and C'_{01} equal to one, and all other independent coefficients in the optimisation process (C_{12}, C'_{03}, and C'_{05}) equal to zero. In contrast to all previous optimisations, this choice of coefficient values no longer corresponds to the *Plate Carrée*, which does not have a pole line half as long as the equator, yet to a doubly symmetric pseudocylindrical map projection of the polynomial type, which is defined by equations (5.8), (5.16) and (5.17), and which has a non-zero value for C_{14} that can be derived from (5.18) (for the *Plate Carrée* C_{14} equals zero). Optimised coefficient values for the pseudocylindrical projection with non-equally-spaced parallels and a pole line half as long as the equator are listed in table 5.9. Figure 5.19 shows the graticule of the projection. Compared to the graticule in figure 5.18 one can see that the shortening of the pole line has somewhat improved the representation of the polar areas. It should be noted that the convergence of the meridians only starts to occur in the higher latitudes. This again indicates that the optimisation process tries to produce close-to-straight angles between the meridians and the parallels for most of the continental area. The N–S stretching of the equatorial areas, already noted in the previous optimisation (figure 5.18), remains.

Table 5.8 Coefficient values and mean finite scale factor for the low-error pseudocylindrical projection with twofold symmetry.

$C_{10} = 0.7920$	$C'_{01} = 1.0304$
$C_{12} = -0.0978$	$C'_{03} = 0.0127$
$C_{14} = 0.0059$	$C'_{05} = -0.0250$
$K_1 = 1.1430$	

Table 5.9 Coefficient values and mean finite scale factor for the low-error pseudocylindrical projection with twofold symmetry and a pole line half the length of the equator.

$C_{10} = 0.7879$	$C'_{01} = 1.0370$
$C_{12} = -0.0238$	$C'_{03} = -0.0059$
$C_{14} = -0.0551$	$C'_{05} = -0.0147$
$K_1 = 1.1453$	

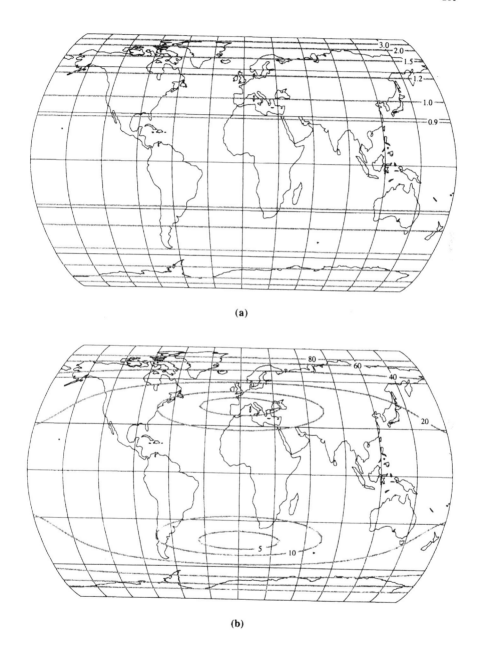

(a)

(b)

Figure 5.18 Low-error pseudocylindrical projection with twofold symmetry, with lines of constant area scale (a), and lines of constant maximum angular distortion (b).

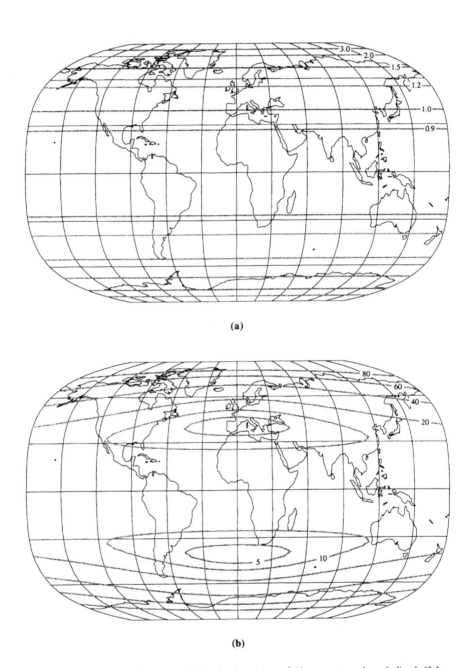

(a)

(b)

Figure 5.19 Low-error pseudocylindrical projection with twofold symmetry and a pole line half the length of the equator, with lines of constant area scale (a), and lines of constant maximum angular distortion (b).

If, instead of obtaining a better representation of the higher latitudes, the mapping of the equatorial areas should be emphasised, one may reduce the distortion of angles in the vicinity of the equator by making the latter twice as long as the central meridian. Imposing a correct ratio of the axes is achieved by putting:

$$C_{10} = C'_{01} + \left(\frac{\pi}{2}\right)^2 \left[C'_{03} + C'_{05} \left(\frac{\pi}{2}\right)^2 \right] \tag{5.19}$$

Optimisation of the remaining five coefficients, starting from the *Plate Carrée*, produces the graticule shown in figure 5.20. The values of the six coefficients that define the projection are listed in table 5.10. The mean distortion value is only slightly increased compared to the original, non-constrained optimisation (table 5.8). Distortion in the equatorial areas is decreased at the cost of a more severe stretching of the polar regions.

Low-error pointed-polar projections

It is well known that world map projections with a pole line of finite length have more balanced distortion patterns then pointed-polar projections (see section 2.5.2). If, for a particular reason, one would like to show the pole as a point anyway, polynomial transformation may help to reduce the compression of the polar areas typical of these projections. One way to derive optimised graticules that show the pole as a point is to apply equations (4.70) and (4.71) to a standard pointed-polar projection. The sinusoidal projection, for example, would be a very good candidate since it combines several useful geometric properties, including symmetry about the equator and the central meridian, equal spacing of the meridians along each parallel, equal spacing of the parallels along the central meridian, and a correct ratio of the axes. A simpler transformation, which expresses the transformed coordinates directly as a function of the longitude and latitude, and offers the same possibilities for the maintenance of geometric properties, is obtained by multiplying each term in (5.9), as well as each term holding the longitude in (5.10) by the cosine of the latitude. Imposing twofold symmetry, an equally divided equator and central meridian, and a correct ratio of the axes, the transformation is defined as:

$$X = R f_1(\lambda, \phi) \cos \phi$$
$$Y = R f_2(\lambda, \phi) \tag{5.20}$$

Table 5.10 Coefficient values and mean finite scale factor for the low-error pseudocylindrical projection with twofold symmetry and a correct ratio of the axes.

$C_{10} = 0.8378$	$C'_{01} = 1.0150$
$C_{12} = -0.1053$	$C'_{03} = 0.0207$
$C_{14} = -0.0011$	$C'_{05} = -0.0375$
$K_1 = 1.1458$	

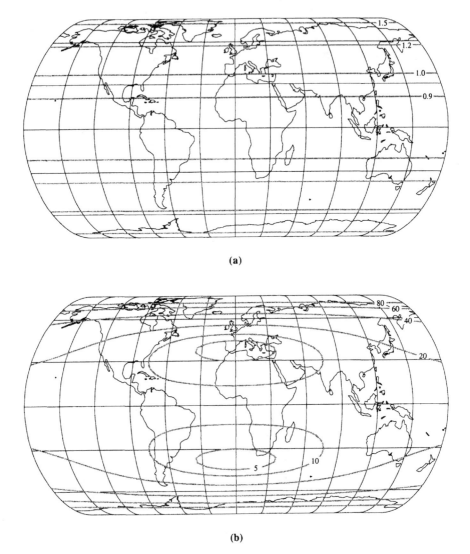

Figure 5.20 Low-error pseudocylindrical projection with twofold symmetry and a correct ratio of the axes, with lines of constant area scale (a), and lines of constant maximum angular distortion (b).

with $f_1(\lambda,\phi)$ as in (5.13) and

$$f_2(\lambda,\phi) = C'_{01}\,\phi + (C'_{21}\,\lambda^2\phi + C'_{41}\,\lambda^4\phi + C'_{23}\,\lambda^2\phi^3)\cos\phi \tag{5.21}$$

and with $C_{10}=C'_{01}$, leaving seven coefficients to be determined. Optimisation (table 5.11, figure 5.21), this time taking Antarctica into account, yields a pointed-polar version of the graticule shown in figure 5.17. Geometric properties are the same for both graticules, only the representation of the pole is different.

As expected, the mean finite scale distortion is significantly higher than for the graticule with pole line (table 5.7). However, compared to the famous Aitoff projection, which has similar geometric characteristics apart from its outline (figure 4.5), the compression in the higher latitudes is reduced by an appropriate adjustment of the curvature of the parallels. The optimised graticule reminds of van der Grinten's fourth projection (1904), his so-called *apfelschnittförmige*, on which the outer meridian is formed by two intersecting circles, and which also has equally spaced axes in a ratio of 2:1 (figure 5.22). Yet the exaggeration of area near the outer meridian is less than on the van der Grinten. Maurer's projection with equidistant parallels and central meridian, presented in 1939, and described in detail in 1944, has a characteristic outline that is even more close to the present projection (Maurer, 1939, 1944). Just like on the van der Grinten IV, however, the parallels are circular arcs, whereas in the present projection the parallels are represented by arbitrary curves.

Optimised pseudocylindrical projections that represent the pole as a point can be derived in a similar way, making $f_1(\lambda,\phi)$ in (5.20) a linear function of the longitude, and $f_2(\lambda,\phi)$ a function of the latitude only. For example, for a doubly symmetric pseudocylindrical map projection with non-equally spaced parallels and a pointed pole, the polynomial model is also given by (5.20), yet with $f_1(\lambda,\phi)$ and $f_2(\lambda,\phi)$ as in (5.16) and (5.17). Imposing a correct ratio of the axes (see (5.19)), and optimising the five remaining coefficients, again taking Antarctica into account, produces a graticule with very favourable distortion characteristics, especially for a pointed-polar projection (figure 5.23, table 5.12). As was already noticed for the optimised pseudocylindrical with shortened pole-line (figure 5.19), also for the optimised pointed-polar projection the convergence of the meridians towards the poles is less sharp than on any of the standard pseudocylindrical projections that have been described in chapter 2. Sharp convergence of the meridians is confined to the higher latitudes, which makes that the general appearance of the projection is not very different from a projection with a pole line, except that there is no discontinuity at the poles. The distortion of angles caused by the projection is moderate for most of the continental area. The area distortion pattern shows two lines of no area distortion in each hemisphere. For the entire zone between the

Table 5.11 Coefficient values and mean finite scale factor for the low-error pointed-polar polyconic projection with twofold symmetry, equally spaced parallels, and a correct ratio of the axes.

$C_{10} = 0.9445$	$C'_{01} = 0.9445$
$C_{12} = 0.1316$	$C'_{21} = 0.0485$
$C_{32} = -0.0145$	$C'_{41} = 0.0063$
$C_{14} = 0.0176$	$C'_{23} = -0.0085$

$K_1 = 1.1861$

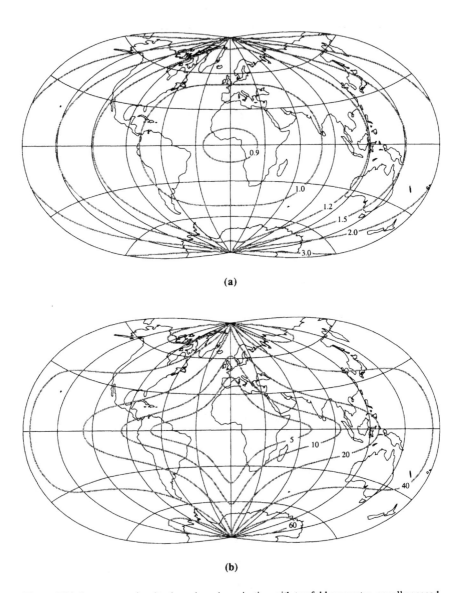

(a)

(b)

Figure 5.21 Low-error pointed-polar polyconic projection with twofold symmetry, equally spaced parallels, and a correct ratio of the axes, with lines of constant area scale (a), and lines of constant maximum angular distortion (b).

parallels of −75° and +75° the areal scale error is between −16% and +13%. This is less than for all optimised projections that have been described so far.

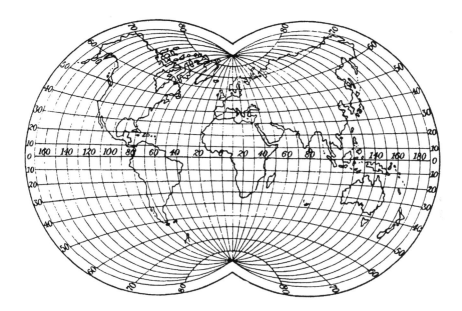

Figure 5.22 van der Grinten's fourth projection (from Snyder, p. 261; © 1993 by The University of Chicago. All rights reserved. Published 1993).

Low-error oblique projections

Changing the aspect of a projection is one of the best ways to reduce distortion for the area to be mapped (see section 2.6). While one may optimise the coordinates of the meta-pole and/or the location of the geographical poles on the meta-graticule for every type of aspect of a standard map projection, and then apply a polynomial transformation to the re-oriented graticule, one may also optimise the aspect of the projection and the polynomial coefficients simultaneously. This will increase the number of parameters to be optimised by one for the normal, the first transverse, and the second transverse aspect, by two for the transverse oblique, the simple oblique, and the skew oblique aspect, and by three for the plagal aspect (table 2.7). We will now present some examples of optimised oblique map projections to

Table 5.12 Coefficient values and mean finite scale factor for the low-error pointed-polar pseudocylindrical projection with twofold symmetry and a correct ratio of the axes.

$C_{10} = 0.8333$	$C'_{01} = 1.0114$
$C_{12} = 0.3385$	$C'_{03} = 0.0243$
$C_{14} = 0.0942$	$C'_{05} = -0.0391$

$K_1 = 1.2151$

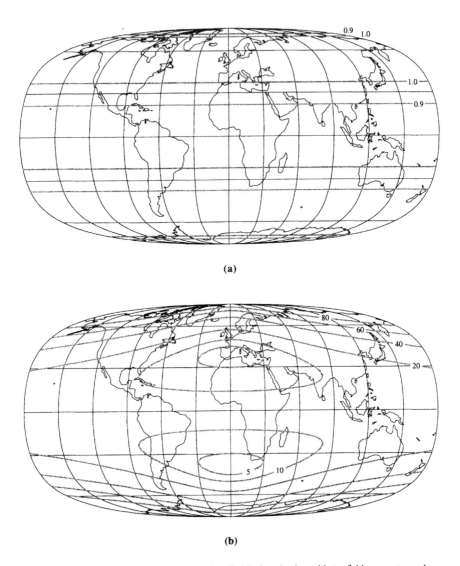

(a)

(b)

Figure 5.23 Low-error pointed-polar pseudocylindrical projection with twofold symmetry and a correct ratio of the axes, with lines of constant area scale (a), and lines of constant maximum angular distortion (b).

illustrate the benefits of the technique, and to draw attention to a few practical considerations related to its use.

First, we will consider the simple oblique aspect. Of all non-conventional aspects, the simple oblique aspect is the most frequently used in commercial mapmaking, usually to portray phenomena that extend beyond one of the poles (e.g.

airlines). The simple oblique aspect has the advantage that it produces a configuration of the continents that, although different from the normal aspect, still looks familiar, and does not confuse the inexperienced map user too much. When choosing for a simple oblique aspect, and applying the polynomial transformation defined by (5.20), (5.13) and (5.21) to the meta-coordinates, a graticule is obtained with centre at 61°N, 24°E (figure 5.24, table 5.13). Although the re-orientation of the graticule leads to an overall distortion that is much lower than for normal aspect optimisations, the southern part of Africa is split up. The split is caused by the high concentration of land in the Northern Hemisphere, which has more weight in the calculation of the overall distortion value, and therefore strongly influences the position of the centre of the projection. Optimisation of other oblique projections, conventional or not, almost invariably leads to the splitting of areas that are located too eccentrically. This demonstrates that the minimisation of overall distortion not necessarily produces useful solutions, and that further adjustments may be required. For the present projection the split of the African continent can easily be avoided by a displacement of the centre of the projection, followed by a re-adjustment of the polynomial coefficients. The optimised graticule that is shown in figure 5.25 is obtained by applying the same constraints as before, only the centre of the projection has been fixed at 45°N, 20°E. Overall distortion is somewhat increased (table 5.14), yet it is much lower than for all normal aspect projections with a regular outline that have been described earlier.

As another example of aspect optimisation the plagal aspect will be considered. Being the most general of all aspects, the plagal aspect offers the highest possible freedom for an optimal positioning of the distortion pattern of the projection, and thus for a maximum reduction of distortion within the area to be

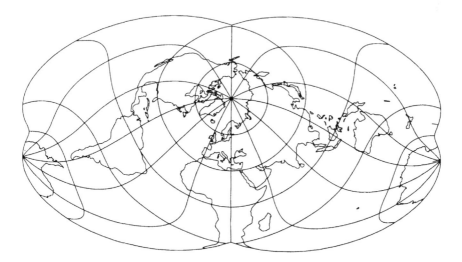

Figure 5.24 Low-error simple oblique polyconic projection with pointed meta-pole and constant scale along the axes.

Table 5.13 Coefficient values and mean finite scale factor for the low-error simple oblique polyconic projection with pointed meta-pole and constant scale along the axes.

$C_{10} = 0.9457$	$C'_{01} = 0.9836$
$C_{12} = 0.2962$	$C'_{21} = 0.0472$
$C_{32} = 0.0420$	$C'_{41} = -0.0106$
$C_{14} = -0.1875$	$C'_{23} = 0.0929$
$K_1 = 1.0583$	

mapped (see section 2.6). Only in the plagal aspect it is possible to show all the continents (including Antarctica) without interruption. In the few cases where plagal aspects of standard map projections have been used, e.g. for the map of the Earth's crust in the *National Geographic Atlas of the World* (NGS, 1995), the absence of interruptions in the continents seems to have been the most important criterion for using this type of representation. Regardless of a rather conservative attitude towards the use of map projections in general (see section 2.6), the main reason why plagal aspects of standard map projections are not used more often in commercial mapmaking is probably the complexity of their graticule, and the highly unfamiliar arrangement of the continents. Some parts of the continental area that occupy a central position in the normal aspect of a projection, and are therefore only moderately distorted on regular world maps, will shift towards the outline of the map in the plagal aspect, and may become severely distorted. This can be considered as a highly unpleasant effect, and may explain why plagal aspects, as well as most other non-conventional aspects, are rarely used.

The extreme distortion of areas for which the shape is familiar to most map readers that is typical for plagal aspects of standard map projections that represent the continents without interruption, can be reduced by using a general polynomial transformation in its plagal aspect, and choosing the coefficients so that the graticule is optimally adapted to the new arrangement of the continental area. To illustrate this we will again optimise the values of the coefficients for the doubly symmetric transformation with equally spaced axes and pointed meta-pole, defined by (5.20), (5.13) and (5.21). Minimisation of the mean finite scale distortion for a set of 5000 random distances covering the entire continental area, Antarctica excluded, produces a graticule with its meta-pole at 29°N, 143°W, and with the meta-longitude of the North Pole equal to 26°E. The graticule, which is not shown here, has the lowest mean finite scale distortion for the continental area of all projections that have been discussed so far, yet Antarctica is slightly interrupted. A small adjustment of the value of the three parameters that define the aspect, followed by optimisation of the polynomial coefficients as before, yields a graticule that is almost identical (table 5.15, figure 5.26), with a mean finite scale distortion that is only slightly higher, and with all the continents shown without interruption. As can be seen, the patterns of distortion are well adapted to the configuration of the continents, with almost for the entire continental area (Antarctica not included) an area scale factor less than 2.0, and a maximum angular distortion less than 40°.

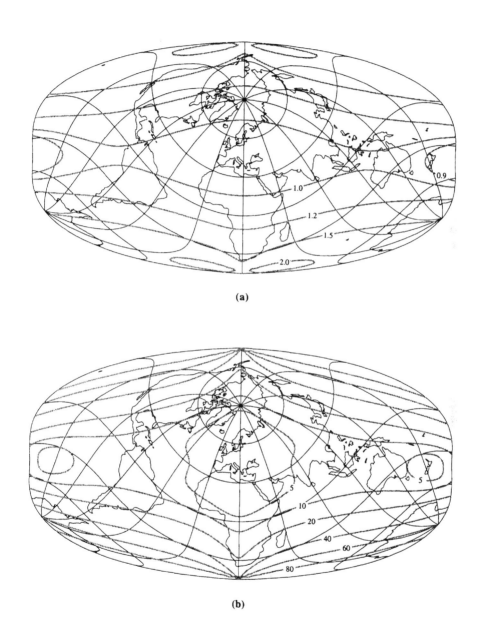

(a)

(b)

Figure 5.25 Low-error simple oblique polyconic projection with pointed meta-pole and constant scale along the axes, centred at 45°N, 20°E, with lines of constant area scale (a), and lines of constant maximum angular distortion (b).

Table 5.14 Coefficient values and mean finite scale factor for the low-error simple oblique polyconic projection with pointed meta-pole and constant scale along the axes, centred at 45°N, 20°E.

$C_{10} = 0.9296$	$C'_{01} = 0.9919$
$C_{12} = 0.2997$	$C'_{21} = 0.0431$
$C_{32} = 0.0200$	$C'_{41} = -0.0064$
$C_{14} = -0.0032$	$C'_{23} = 0.0273$
$K_1 = 1.0737$	

Table 5.15 Coefficient values and mean finite scale factor for the low-error plagal aspect polyconic projection with pointed meta-pole (30°N, 140°W), geographical North Pole at a meta-longitude of 30°, and constant scale along the axes.

$C_{10} = 0.9523$	$C'_{01} = 0.9774$
$C_{12} = 0.2913$	$C'_{21} = 0.0281$
$C_{32} = 0.0495$	$C'_{41} = -0.0054$
$C_{14} = -0.2105$	$C'_{23} = 0.1293$
$K_1 = 1.0506$	

5.3.3 Low-error equal-area projections

In section 4.2.2, two polynomial transformations have been proposed that satisfy the general condition for an equal-area transformation. Applied to an equal-area projection they both produce a new equal-area graticule that maintains some of the geometric properties of the original projection. Combining both transformations into one single transformation increases the flexibility for graticule adjustment, offering the mapmaker the possibility to decide which geometric properties he wishes to maintain, and which he does not. For example, using fifth-order polynomials, and imposing symmetry about both the x- and the y-axis, the combined transformation defined by (4.68) reduces to:

$$X = C_1 x + C_3 x^3 + C_5 x^5$$
$$Y = y/(C_1 + 3C_3 x^2 + 5C_5 x^4) \tag{5.22}$$

followed by

$$X' = X/(C'_1 + 3C'_3 Y^2 + 5C'_5 Y^4)$$
$$Y' = C'_1 Y + C'_3 Y^3 + C'_5 Y^5 \tag{5.23}$$

with six coefficients to be optimised. Maintenance of the correct ratio of the axes

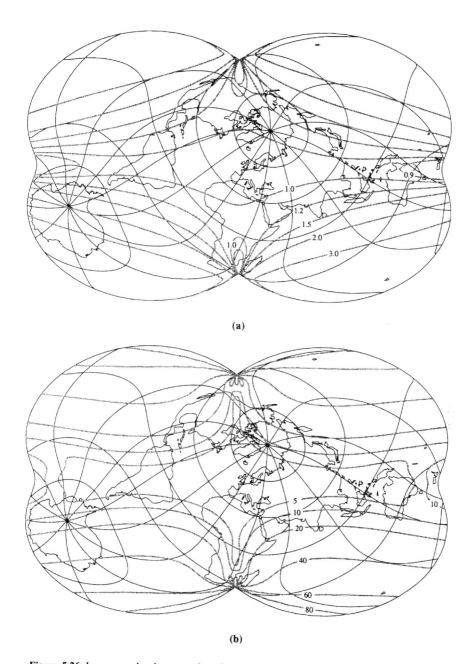

(a)

(b)

Figure 5.26 Low-error plagal aspect polyconic projection with pointed meta-pole (30°N, 140°W), geographical North Pole at a meta-longitude of 30°, and constant scale along the axes, with lines of constant area scale (a), and lines of constant maximum angular distortion (b).

can be imposed in a similar way as explained for general polynomial transformations (see section 4.1.3). Starting with the first transformation (equation (5.22)), it can be shown that the maintenance of the correct ratio of the axes is obtained by putting

$$C_1 = \frac{-K_1 + \sqrt{K_1^2 + 4x_E^2}}{2x_E} \tag{5.24}$$

with x_E the maximum value of the x-coordinate on the equator ($\lambda = \pi$, $\phi = 0$) of the parent projection, and

$$K_1 = C_3 x_E^3 + C_5 x_E^5 \tag{5.25}$$

For the second transformation (equation (5.23)) the maintenance of the correct ratio is obtained by putting

$$C_1' = \frac{-K_2 + \sqrt{K_2^2 + 4Y_P^2}}{2Y_P} \tag{5.26}$$

with Y_P the maximum value of the Y-coordinate on the central meridian ($\lambda = 0$, $\phi = \pi/2$) of the graticule obtained from the first transformation, and

$$K_2 = C_3' Y_P^3 + C_5' Y_P^5 \tag{5.27}$$

As can be seen, imposing a correct ratio of the axes reduces the number of coefficients to be optimised by two.

Projections with curved parallels

Applying (5.22) and (5.23) to the Hammer–Wagner projection, which has the lowest mean finite scale distortion of all standard equal-area projections that have been evaluated in this study (table 3.5), and maintaining the 2:1 ratio of the axes, produces the graticule shown in figure 5.27. Polynomial coefficients and mean finite scale distortion are listed in table 5.16. The N–S stretching of the continents, which is typical of equal-area graticules with a pole line, and which is well noticed on the parent projection (figure 4.8), is less prominent on the optimised graticule. However, the overall improvement in the appearance of the continents is clearly at the cost of those areas that are located most eccentrically (polar regions, areas close to the outer meridian).

What is probably most disturbing about the graticule in figure 5.27 is the strong variation of scale along the equator, which is caused by the first transformation (equation (5.22)). Applying only the second transformation, which maintains the relative spacing of the parallels along the equator as it is defined in the original projection, and optimising the values of the three coefficients, yields

Table 5.16 Coefficient values and mean finite scale factor for the low-error equal-area transformation of Hammer–Wagner with twofold symmetry and a correct ratio of the axes.

$C_1 = 1.0706$	$C'_1 = 1.0090$
$C_3 = 0.0000$	$C'_3 = -0.0030$
$C_5 = -0.0027$	$C'_5 = -0.0055$
$K_1 = 1.1510$	

the graticule shown in figure 5.28 (coefficients are listed in table 5.17). What is most striking about this graticule is the moderate curvature of the parallels, which makes it look almost like a pseudocylindrical projection. The graticule has a rather long pole line, comparable with the low-error pseudocylindricals that have been discussed in the previous section. While this avoids the presence of sharp intersections of meridians and parallels in the peripheral parts of the map, it also introduces a stronger variation of scale along the central meridian, and a more severe E–W stretching of the polar areas, two well-known effects of the use of a pole line that are also observed on standard equal-area projections (see section 2.5.2).

To obtain an optimised equal-area projection with less E–W stretching in the polar areas the second transformation (equation (5.23)) was applied to the Hammer–Aitoff projection. Doing so produces an equal-area projection that represents the pole as a point. The low-error graticule, obtained by minimisation of distortion for the entire continental area (Antarctica included), is shown in figure

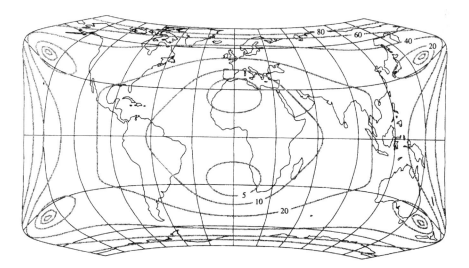

Figure 5.27 Low-error equal-area transformation of Hammer–Wagner with twofold symmetry and a correct ratio of the axes, with lines of constant maximum angular distortion.

Table 5.17 Coefficient values and mean finite scale factor for the low-error equal-area transformation of Hammer–Wagner with twofold symmetry and constant scale along the equator.

$C'_1 = 1.0099$
$C'_3 = -0.0227$
$C'_5 = -0.0051$
$K_1 = 1.1588$

5.29. Its polynomial coefficients are listed in table 5.18. Although the equal-area property and the representation of the pole as a single point are not considered the most successful design options for the development of world maps, one can see that the graticule has a quite favourable distortion pattern for a projection of this type. Compared to most standard equal-area projections that represent the pole as a point, it causes less shearing of areas that are far away from the centre of the projection. On the other hand, it avoids the strong E–W stretching which is typical of equal-area projections that represent the pole as a line. High distortion values are confined to the upper NW and NE corners of the map. Like in the original Hammer–Aitoff projection, the polar areas within the inner hemisphere are mapped with much less distortion than in any pseudocylindrical equal-area projection.

Projections with straight parallels

As explained in the introduction to this study, it was especially in the beginning of the 20th century that many new pseudocylindrical equal-area projections were

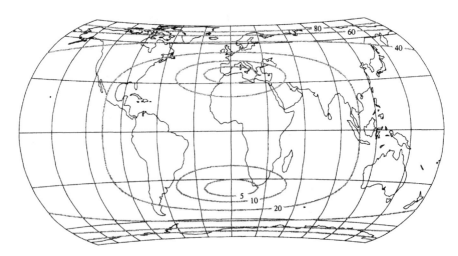

Figure 5.28 Low-error equal-area transformation of Hammer–Wagner with twofold symmetry and constant scale along the equator, with lines of constant maximum angular distortion.

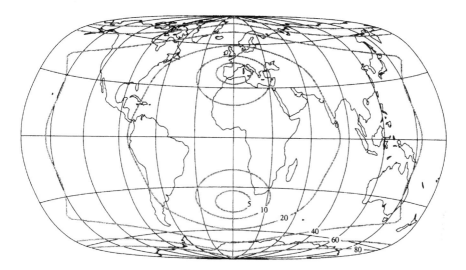

Figure 5.29 Low-error equal-area transformation of Hammer–Aitoff with twofold symmetry and constant scale along the equator, with lines of constant maximum angular distortion.

Table 5.18 Coefficient values and mean finite scale factor for the low-error equal-area transformation of Hammer–Aitoff with twofold symmetry and constant scale along the equator.

$C'_1 = 1.1145$
$C'_3 = -0.0185$
$C'_5 = -0.0201$
$K_1 = 1.2071$

developed. Some of them remained academic presentations, yet others became quite famous, and are still found in contemporary atlases, although today often in interrupted form, or in an oblique aspect. The former success of these projections probably lies in the combination of the equal-area property, which was for a very long time considered as the most relevant property for a world map, with a very simple, and for that reason very attractive graticule geometry. The main obstacle in developing a pseudocylindrical equal-area projection is to obtain a good balance between the spacing of the parallels (avoiding strong variation in scale along the central meridian) and the curvature of the meridians (avoiding sharp intersections of the graticule or E–W stretching in the higher latitudes). Since polynomial optimisation offers more flexibility for adjustment of graticule characteristics than more conventional approaches to map projection design, it is an interesting technique to explore this balancing of seemingly incompatible goals, especially in those cases where it has been proven difficult to find an acceptable compromise.

For the second of the two equal-area transformations that have been described above (equation (5.23)), the X is a linear function of the x-coordinate of the parent projection, while the Y is a function of the y-coordinate only. Therefore this second transformation can be applied to derive low-error pseudocylindrical equal-area projections from standard cylindrical or pseudocylindrical equal-area graticules. Of course, as before, the choice of the parent graticule will depend on the geometric features that should be present in the optimised projection. Since all standard pseudocylindrical equal-area projections that represent the pole as a line stretch the polar areas and elongate the N–S distances in the lower latitudes, we will develop a low-error equal-area projection that represents the pole as a point. To do so we will start with the sinusoidal projection, apply equation (5.23), and optimise the transformation coefficients in the hope that the strong curvature of the meridians on the sinusoidal projection will be modified in such a way that a good balance is obtained in the representation of the higher and the lower latitudes. By maintaining a correct ratio of the axes (equation (5.26), (5.27)) the number of parameters to be optimised will be reduced to two, and the impact of the optimisation will be restricted to a modification of the shape of the outer meridian.

Optimisation of the coefficients in the transformation for the continental area, Antarctica not included, produces the map shown in figure 5.30. The values of the three coefficients are found in table 5.19. What is most remarkable about this projection is the seeming presence of a pole line, which is not really there. Indeed, since the optimisation process starts with the sinusoidal projection all meridians in the optimised graticule do meet in one point, only the shape of the meridians has been modified in such a way that it seems as if the graticule really has a pole line. This can be interpreted as the ultimate justification for the introduction of a pole line, as was done by many developers of pseudocylindrical map projections in the

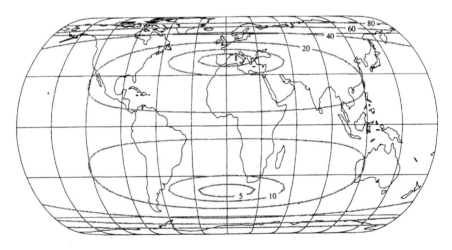

Figure 5.30 Low-error equal-area transformation of the sinusoidal projection with twofold symmetry, equally divided, straight parallels, and a correct ratio of the axes, with lines of constant maximum angular distortion (not including Antarctica in the optimisation).

Table 5.19 Coefficient values and mean finite scale factor for the low-error equal-area transformation of the sinusoidal projection with twofold symmetry, equally divided, straight parallels, and a correct ratio of the axes (not including Antarctica in the optimisation).

$$C'_1 = 1.1988$$
$$C'_3 = -0.1290$$
$$C'_5 = -0.0076$$

$$K_1 = 1.1676$$

Table 5.20 Coefficient values and mean finite scale factor for the low-error equal-area transformation of the sinusoidal projection with twofold symmetry, equally divided, straight parallels, and a correct ratio of the axes (including Antarctica in the optimisation).

$$C'_1 = 1.1481$$
$$C'_3 = -0.0753$$
$$C'_5 = -0.0150$$

$$K_1 = 1.2250$$

last century. Unfortunately, the optimised projection also suffers from all deficiencies of pseudocylindricals with a pole line that have been discussed above. In fact, the optimised projection in figure 5.30 shows some remarkable similarities with Eckert's fourth projection (figure 2.11, see also Canters and Decleir 1989, p. 119), which is the graticule with the least finite scale distortion of all standard pseudocylindrical equal-area projections that have been discussed in this study (see table 3.5). While the distortion patterns for the two projections are slightly different, the mean finite distortion values for both projections are almost equal.

The emerging of a virtual pole line, as described above; is a phenomenon that not only occurs in the optimisation of equal-area pseudocylindricals. It is observed for almost any normal aspect optimisation of a pointed-polar projection of the polynomial type and is the main reason why Antarctica has been included in all pointed-polar optimisations that have been discussed earlier in this chapter. Repeating the previous optimisation, yet this time with Antarctica added to the area for which distortion is minimised, leads to the projection that is shown in figure 5.31. The three polynomial coefficients and the mean distortion value are listed in table 5.20. Although the graticule has more distortion in its NW, NE, SW and SE parts than the previous one, the polar and equatorial areas are represented with less distortion. The curvature of the meridians is such that the projection holds the middle between the typical pointed-polar pseudocylindrical projection and a projection that represents the pole as a line, introducing sharp intersections in the graticule only in the immediate vicinity of the pole and keeping scale variation along the central meridian within acceptable bounds. The "delayed" convergence of the meridians creates a pleasant perspective effect that is

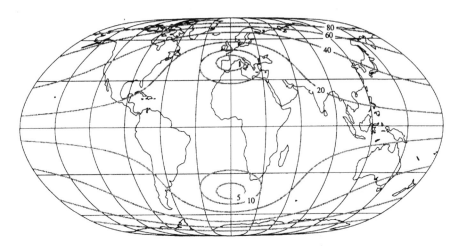

Figure 5.31 Low-error equal-area transformation of the sinusoidal projection with twofold symmetry, equally divided, straight parallels, and a correct ratio of the axes, with lines of constant maximum angular distortion (including Antarctica in the optimisation).

also recognised in the optimised version of the non-equal-area pointed-polar pseudocylindrical projection (figure 5.23).

5.4   LOW-ERROR PROJECTIONS FOR CONTINENTAL AND REGIONAL MAPS

The use of finite distortion criteria, for the development of optimal map projections, is not restricted to global mapping, although the advantages of finite distortion measurement, as compared to the more traditional, infinitesimal approach to error minimisation, are expected to be less pronounced when the size of the region to be mapped becomes smaller. If visible distortion is the only concern, as it is often the case in small-scale mapping, then optimisation of the parameters of standard projections will mostly produce satisfactory results. However, the optimisation of projection parameters can always be followed by polynomial transformation, if a further reduction of error should be required. This may be so for the mapping of large parts of the Earth's surface, if it proves that the periphery of the map still shows severe distortion after projection parameters have been optimised. The various transformations that have been discussed for global mapping may also be applied for mapping at the continental scale, although some of the constraints that have been formulated for the development of world maps may no longer be relevant (e.g. symmetry about the equator, correct ratio of the axes, ...).

The need for an additional reduction of distortion, once standard projection parameters have been optimised, may also occur for smaller areas (regional mapping) if the outline of the region to be mapped is very irregular and does not fit the distortion pattern of any standard projection. Polynomial transformation may then lead to a substantial reduction of the overall error and the range of scale factors, as was clearly demonstrated in Snyder's work on the development of a low-error conformal map for the United States, including Alaska and Hawaii (Snyder, 1984a) (figure 4.14). Although Snyder minimises the root-mean-square of local scale errors, calculated for a set of points in the area, minimisation of finite distortion, as proposed in this study, is expected to produce similar improvements. So far the use of polynomials in minimum-error research has been mostly limited to conformal mapping. For most small-scale applications, however, the equal-area property is more important. The double equal-area polynomial transformation that has been proposed in chapter 4, and that has been applied for global mapping purposes in the previous section, may be very useful for the development of equal-area maps with less distortion at the regional scale. This will be shown in the following examples.

To demonstrate the use of the finite distortion approach at the continental and sub-continental scale, three areas have been selected. In the first case study (North and South America) finite distortion will be minimised to determine the optimal aspect, and the optimal position of the standard lines for different projections of the conic group. The second study addresses the development of an equal-area counterpart to Miller's low-error conformal map projection for Europe and Africa (Miller, 1953). In the last study attention will be focused on the development of a low-error equal-area graticule for the fifteen member states of the European Union. In all examples, the calculation of the mean finite scale distortion will be based on a sample of 5000 distances, randomly selected over the area to be mapped.

5.4.1 Optimisation of projection parameters for maps of North and South America

When selecting a projection for the mapping of a particular area, the choice of aspect is extremely important. An optimal positioning of the point or line(s) of zero distortion with respect to the Earth's surface will put the region to be mapped in the least distorted part of the map projection, and will guarantee minimum overall distortion or minimum variation of scale within the map area, depending on the criterion for optimisation that is used. If only conic type projections are considered, the choice of projection will depend on the shape of the region. Regions that are approximately circular in outline are usually represented in an azimuthal projection, whereas elongated regions are mapped on a conical or cylindrical projection, depending on whether the longer axis through the region is best approximated by a small or by a great circle (see section 6.1). In practice, however, this choice is not always easy to make. Not all countries can be designated as predominantly round or elongated along one axis and, if an oblique axis can be identified, it is often not easy to decide if it best fits a great or a small circle by simply looking at a globe. The best way to find out which projection is to be preferred is by optimising the aspect for each of them, and comparing overall distortion values or maximum scale variation. The need for aspect optimisation and objective comparison of distortion values is even greater if other than conic type projections are considered as well.

Table 5.21 Parameter values and mean finite scale factor for nine low-error conic type map projections for North and South America.

	Equidistant	Equal-area	Conformal
Oblique azimuthal	$\phi_p = 26.39°$	$\phi_p = 25.55°$	$\phi_p = 28.23°$
	$\lambda_p = -69.32°$	$\lambda_p = -74.33°$	$\lambda_p = -58.51°$
	$K_1 = 1.0317$	$K_1 = 1.0398$	$K_1 = 1.0593$
Oblique cylindrical	$\phi_p = -22.13°$	$\phi_p = -21.52°$	$\phi_p = -21.48°$
	$\lambda_p = -157.70°$	$\lambda_p = -157.28°$	$\lambda_p = -156.86°$
	$\phi'_0 = 7.37°$	$\phi'_0 = 7.98°$	
	$K_1 = 1.0071$	$K_1 = 1.0086$	$K_1 = 1.0134$
Oblique conical	$\phi_p = 30.33°$	$\phi_p = 29.77°$	$\phi_p = 29.35°$
	$\lambda_p = -8.76°$	$\lambda_p = -8.84°$	$\lambda_p = -7.61°$
	$\phi'_1 = 15.96°$	$\phi'_1 = 15.70°$	$\phi'_1 = 10.36°$
	$\phi'_2 = 31.35°$	$\phi'_2 = 32.02°$	$\phi'_2 = 34.16°$
	$K_1 = 1.0061$	$K_1 = 1.0073$	$K_1 = 1.0111$

ϕ_p, λ_p: latitude and longitude of the meta-pole

ϕ'_0, ϕ'_1, ϕ'_2: meta-latitude of the standard line(s)

Limiting our choice to projections of the conic type, it is clear that North and South America together are best mapped on a cylindrical or conical map projection. Since both continents form an area with a rather elongated shape, azimuthal projections are expected to produce a large variation in scale along the major axis of the area. The landmasses of both continents roughly extent in a NW–SE direction, thus an oblique aspect seems the most obvious choice. Careful examination of the position of both continents on the globe shows that if one tries to fit an oblique great circle through the area North America will be located mostly east of this circle, while South America will be located mostly west of it. Hence one might expect an oblique small circle to provide a better fit. If this is so, the use of a conical projection might lead to a substantial reduction of distortion in comparison with the use of a cylindrical projection. To determine optimal aspects for both projections, and to find out if the gain in accuracy that is obtained by using a conical instead of a cylindrical projection is significant, overall finite distortion can be minimised.

Table 5.21 lists meta-pole coordinates, position of the standard lines, and mean finite scale distortion for the optimal aspect of the three common types of cylindrical and conical projections. Optimal aspects for the three corresponding azimuthal projections are also given, just to show that for the case of North and South America overall distortion for these projections is much higher than for the cylindrical and conical projections, as can be inferred by the shape of the area. For the cylindrical projections the meta-pole is situated in the South Pacific. The meta-equator is thus NW–SE oriented. It makes an angle of about 22° (ϕ_p) with the polar axis, and cuts the true equator at a longitude between $-65°$ and $-70°$ ($\lambda_p + 90°$). For the conical projections the meta-pole is located north of the equator, in South

Marocco. Both standard lines intersect the North and South American continent and are about 15° apart for the equidistant and equal-area version, 24° for the conformal version. As can be seen from table 5.21, the mean finite distortion values for the conical projections are only slightly lower than for their cylindrical counterparts. Using finite distortion as a decisive criterion both types of projections prove to be almost equally suited. Because the range in meta-latitude is limited, distortions for equidistant and equal-area projections are very alike. For conformal projections much higher distortion values are obtained.

As was already pointed out in the discussion about function minimisation (see section 5.2), the optimal position of the meta-pole can only be found by making a proper initial guess. In the case of the mapping of North and South America, optimisation of the aspect of the conical and cylindrical projection produces two solutions, dependent on the initial position of the meta-pole. When chosen in the Atlantic, the solution listed in table 5.21 is obtained. When chosen in the Pacific, another solution is found. For the cylindrical projections the second solution proves to be identical to the first (same meta-equator, same position of the standard lines), yet the "north" and the "south" pole are exchanged. For the conical projections new meta-poles are obtained that are located in the South Pacific, close to French Polynesia. Mean distortion values for these projections are higher than for their Atlantic counterparts, but still lower than for the optimal cylindrical projections.

Figure 5.32 shows optimal aspects for both the cylindrical equal-area and the conical equal-area projection, with lines of constant maximum scale error superimposed. The distortion of North and South America is below the threshold of visibility, yet Greenland, which is located most eccentrically, looks quite different on both projections. This confirms the known fact that one should be careful in using cylindrical and conical projections for the mapping of areas with large meta-latitudinal range, especially if the equal-area property is required. On the other hand, the results also prove that cylindrical and conical projections can be used with satisfaction for the mapping of relatively large areas if the aspect of the projection can be chosen freely, and the ultimate concern is to avoid visible distortion. While the oblique aspect of the azimuthal projection is well-known, oblique aspects of cylindrical and conical projections are rarely found in atlases, although Hammer already discussed the merits of using oblique aspect conical projections a century ago (Hammer, 1889b). The only well-known example of the use of an oblique aspect conical projection for small-scale mapping purposes is Miller's bipolar oblique conical conformal projection, also designed for small-scale maps of North and South America (Miller, 1941). To accommodate the fact that both continents curve in opposite directions as one proceeds from north to south (see above), Miller used separate conical projections for each continent. Both meta-poles are located on opposite sides of a great circle arc, oriented SW–NE and passing through Central America. This great circle also limits the area that is covered by each projection. Standard lines for both projections were chosen to match at the juncture. For other positions along the divide there is a small discontinuity that is resolved by local adjustment.

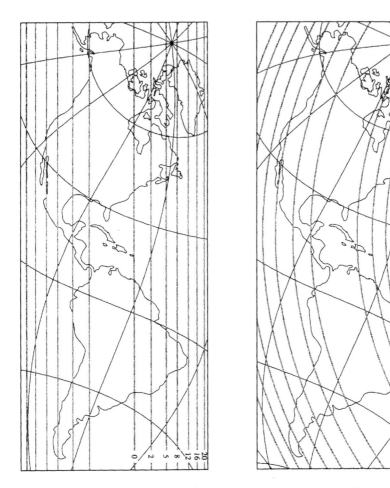

Figure 5.32 Low-error oblique cylindrical equal-area projection (left), and low-error oblique conical equal-area projection (right) for maps of North and South America, with lines of constant maximum scale error (%).

5.4.2   Tailor-made equal-area projections for Europe and Africa

The shape and orientation of Europe and Africa together suggests that both continents are best represented by means of an oblique azimuthal or a transverse cylindrical projection. Which of both projections is to be preferred is difficult to decide. The area is not circular in outline, nor is it narrowly stretched along its central meridian. Its shape is more oval-like, the longer axis coinciding with the meridian of 20°, and roughly extending from − 40° to 80° in latitude, the smaller axis close to the parallel of 20° and extending from − 20° to 60° in longitude. Hence none of both candidate projections will be an ideal choice. The use of an oblique azimuthal projection will lead to a relatively high variation in scale along the central meridian, while a transverse cylindrical projection will produce high distortion values along the east and west extremities of the map. Optimising the parameters for the equal-area version of both projections, rounding off the coordinates of the meta-pole to the nearest five degrees, shows that the two have very similar mean finite distortion values (table 5.22), although the range of scale factors is significantly higher for the transverse cylindrical (figure 5.33). Previous work by Miller (1953) and Tobler (1974) suggests that better results can be obtained by transforming the oblique azimuthal projection so that isocols are better adjusted to the typical shape of the area.

In 1953, Miller developed a new low-error conformal projection for a general-purpose map of Europe and Africa on the scale of 1 : 5 000 000. Although an equal-area projection might be considered more desirable for so large an area, the decision to develop a conformal projection was based on the fact that manual plotting from original source material on larger scales is much easier and accurate if map detail is transformed from one conformal projection to another, rather than to an equal-area or other type of projection (Miller, 1953). To limit scale departures and area distortion, Miller applied a third-order complex-algebra polynomial transformation to an oblique aspect stereographic projection of the area centred at 18°N, 20°E (see section 4.2.3). The result is a projection with oval instead of circular isocols, with their major axis on the central meridian (figure 4.13). By

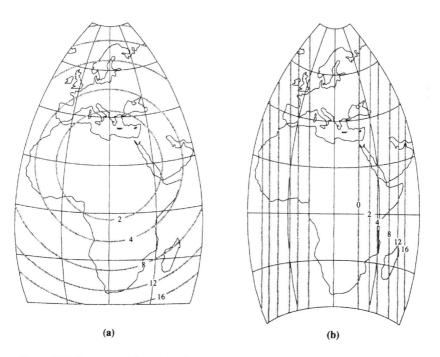

(a)   **(b)**

Figure 5.33 Low-error oblique azimuthal equal-area projection (a), and low-error transverse cylindrical equal-area projection (b) for Eurafrica, with lines of constant maximum scale error (%).

Table 5.22 Parameter values and mean finite scale factor for optimised versions of the oblique azimuthal equal-area projection and the transverse cylindrical equal-area projection for Eurafrica.

	Parameter values	Mean finite scale factor
Oblique azimuthal equal-area	$\phi_p = 20°$ $\lambda_p = 20°$	1.0163
Transverse cylindrical equal-area	$\phi_p = 0°$ $\lambda_p = -70°$ $\phi'_0 = 10.6°$	1.0150

ϕ_p, λ_p: latitude and longitude of the meta-pole

ϕ'_0: meta-latitude of the standard lines

shaping the isocols so that they better follow the general outline of the area the graticule comes closer to the theoretical minimum-error case, which is obtained if the area is completely surrounded by a line of constant scale (Chebyshev's theorem).

With digital mapping methods, there are no difficulties in transforming map detail from existing maps to a common equal-area reference system, if the projection used for each of the base maps is fully described. Equal-area counterparts to Miller's low-error conformal map can be derived by applying polynomial transformations of the type discussed in section 4.2.2 to a standard oblique equal-area projection centred on the area, and then optimising the values of the polynomial coefficients. Already in 1974, Tobler proposed a simple linear (first-order) transformation of the oblique azimuthal equal-area projection that produces equal maximum scale factors at the extreme points along the two major axes of the area to be mapped (for Eurafrica the extreme points were chosen 38° from the centre on the x-axis, and 52° from the centre on the y-axis) (figure 5.34a) (for more details, see section 4.2.2). Although the transformation minimises the range of scale factors, it should be noticed that the mean distortion of finite distances is higher than for the standard oblique azimuthal projection, which has a better distribution of scale error close to the centre of the area. Applying Tobler's first-order transformation, but determining the affine coefficient by minimisation of the mean distortion of finite distances produces another result that does not differ much from the original projection (table 5.23, figure 5.34b).

To see what can be attained with higher-order transformations, third-order and fifth-order double equal-area transformations, as defined by equations (4.68), have been applied to the oblique azimuthal equal-area projection with centre at 20°N, 20°E, and this for various symmetry conditions (twofold symmetry, one-fold symmetry, no symmetry). Table 5.23 summarises optimal coefficient values and mean distortion of finite distances for each of the six polynomial transformations. Figures 5.35, 5.36 and 5.37 show the corresponding distributions of maximum scale error for the twofold symmetric, one-fold symmetric and asymmetric case. Examination of table 5.23 indicates that the use of higher-order transformations

Table 5.23 Parameter values and mean finite scale factor for various low-error equal-area transformations of the oblique azimuthal equal-area projection, for the continuous mapping of Europe and Africa.

	Parameter values		Mean finite scale factor
FIRST-ORDER TRANSFORMATION (*)			
with equal maximum scale factor along the two major axes	$C_1 = 0.9750$	$C'_1 = 1.0256$	1.0206
with minimum mean finite scale factor	$C_1 = 0.9969$	$C'_1 = 1.0031$	1.0161
THIRD-ORDER TRANSFORMATION (*)			
with twofold symmetry	$C_1 = 1.0061$	$C'_1 = 0.9966$	1.0126
	$C_3 = -0.0583$	$C'_3 = 0.0435$	
with one-fold symmetry	$C_1 = 1.0162$	$C'_1 = 1.0061$	1.0125
	$C_3 = -0.0619$	$C'_2 = -0.0057$	
		$C'_3 = 0.0450$	
without symmetry	$C_1 = 1.0175$	$C'_1 = 1.0079$	1.0123
	$C_2 = -0.0069$	$C'_2 = -0.0057$	
	$C_3 = -0.0542$	$C'_3 = 0.0434$	
FIFTH-ORDER TRANSFORMATION (*)			
with twofold symmetry	$C_1 = 1.0058$	$C'_1 = 0.9948$	1.0126
	$C_3 = -0.0810$	$C'_3 = 0.0442$	
	$C_5 = 0.0873$	$C'_5 = 0.0008$	
with one-fold symmetry	$C_1 = 1.0182$	$C'_1 = 1.0058$	1.0123
	$C_3 = -0.0918$	$C'_2 = -0.0056$	
	$C_5 = 0.1128$	$C'_3 = 0.0494$	
		$C'_4 = -0.0013$	
		$C'_5 = -0.0042$	
without symmetry	$C_1 = 1.0048$	$C'_1 = 0.9962$	1.0117
	$C_2 = 0.0041$	$C'_2 = -0.0054$	
	$C_3 = -0.0745$	$C'_3 = 0.0340$	
	$C_4 = -0.0739$	$C'_4 = -0.0008$	
	$C_5 = 0.1097$	$C'_5 = 0.0087$	

(*) Parent projection: oblique azimuthal equal-area projection with meta-pole at 20°N, 20°E

reduces the mean scale error to ±75% of its original value. The use of fifth-order instead of third-order polynomials does not seem to have a major impact on the mean scale error, at least not for this example. The same can be said about the use of different symmetry conditions. In terms of the observed range of scale errors, however, fifth-order transformations are clearly superior. A comparison of distortion patterns shows that isocols for fifth-order transformations are more closely adapted to the general shape of the area. As a result, scale error for most points along the outer boundaries of Eurafrica is less extreme. Lowering the level of symmetry has a similar impact on the distribution of error.

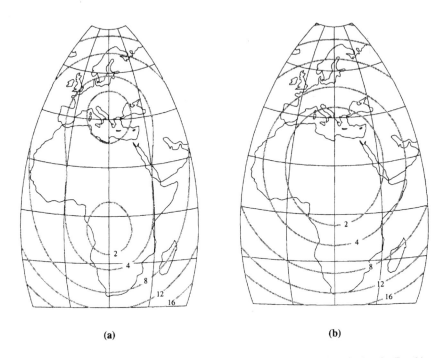

(a)   (b)

Figure 5.34 First-order transformation of the oblique azimuthal equal-area projection for Eurafrica: with equal maximum scale factors along the two major axes of the area (Tobler, 1974) (a), with minimum finite scale distortion (b) (with lines of constant maximum scale error (%)).

In spite of these general observations, one should be very cautious with an increase of polynomial order or with the release of symmetry conditions. While both actions leave more flexibility for graticule adjustment, they also increase the risk of reducing the mean distortion value for the entire region at the cost of an increase of distortion in one or more peripheral areas. In the present example, the use of a fifth-order, instead of a third-order, polynomial transformation clearly increases distortion for the Isle of Malagasia. For the fifth-order transformation with one-fold symmetry maximum scale error at the outer edge of Malagasia reaches 16%, which is more than the extreme scale error for the entire area on the standard oblique azimuthal equal-area projection (12%). In other words, although the use of higher-order polynomial transformations with optimal coefficients tends to reduce overall scale error in the outer parts of the mapped area, it does not necessarily reduce the extreme scale error value for the area. The problem of scale variance was already mentioned earlier, when we were discussing the development of low-error world maps. An uneven distribution of scale error in the outer parts of an area will mostly occur if the area has an irregular shape. It can only be resolved by developing low-error maps based on optimisation criteria that specify a minimum error variance *along* a region's boundary, instead of a minimum overall error *within* its boundary (Nestorov, 1997; see also section 5.1).

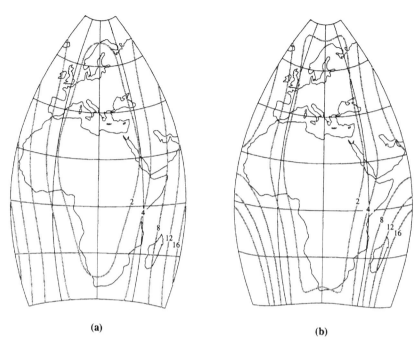

(a) **(b)**

Figure 5.35 Twofold symmetric low-error transformation of the oblique azimuthal equal-area projection for Eurafrica: third-order transformation (a), fifth-order transformation (b) (with lines of constant maximum scale error (%)).

5.4.3 Low-error equal-area projections for the European Union

The problem of map projection choice is not restricted to traditional map production. The growing importance of geographical information systems and the establishment of spatial databases at the regional and the global scale has promoted new interest in the subject of map projection selection (Clark *et al.*, 1991; ESF, 1992; Canters, 1992, 1995, see also chapter 6). In contrast to the paper map, digital databases are not physically restricted to a two-dimensional representation of the Earth's surface. Storing data by their geographical coordinates therefore seems the ideal solution, at least theoretically, since it allows cartometric operations to be performed with negligible error. Unfortunately, most GIS software operates in two dimensions. Hence the traditional problem of map projection distortion remains.

While for most applications at the local up to the national level the reference system that is used for the official topographic mapping of a country's territory can be adopted, the problem of map projection selection will usually present itself when data coming from different countries have to be integrated into a common reference system. The demand for a continuous coverage of a large area, formed by

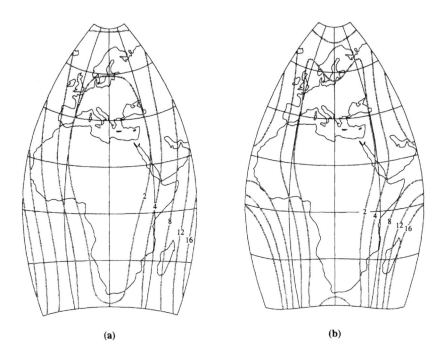

(a)   (b)

Figure 5.36 One-fold symmetric low-error transformation of the oblique azimuthal equal-area projection for Eurafrica: third-order transformation (a), fifth-order transformation (b) (with lines of constant maximum scale error (%)).

various adjacent or non-adjacent countries, may lead to a substantial geometrical distortion near the extremities of the area. As such, the use of standard map projections may be insufficient to meet the error conditions one is willing to accept. Depending on the shape of the area to be covered, error-reducing transformations, as previously described, may lower the range of distortion values substantially.

Spatial data integration is a key issue in the development of geographical databases for the European Union (EU) (Briggs and Mounsey, 1989). Some years ago, Derek Maling was asked to propose a suitable map projection for *CORINE* (Co-ordinated Information on the European Environment), the EU's environmental database (Maling 1992, pp. 256–62). In view of the actual use of *CORINE*, an equal-area projection was considered the most appropriate for storage and continuous mapping of the data. Using a combination of graphical and analytical methods (overlay of isocols on an atlas map of the area, and calculation of distortion values for selected points), Maling determined the optimal aspect and other projection parameters for various standard equal-area projections, and compared distortion characteristics for the limiting extremities on each projection.

Since the EU at that time consisted of twelve member states, extending in

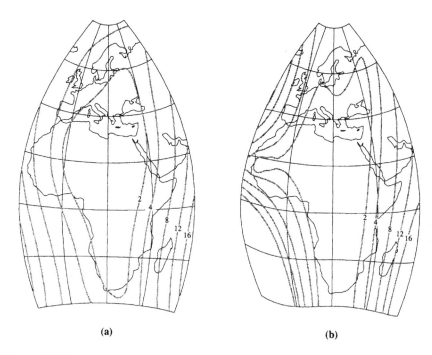

(a)   (b)

Figure 5.37 Non-symmetric low-error transformation of the oblique azimuthal equal-area projection for Eurafrica: third-order transformation (a), fifth-order transformation (b) (with lines of constant maximum scale error (%)).

two major, nearly perpendicular directions, which Maling called the Belfast – Alexandria arc (NW–SE) and the Cape St-Vincent – Gdansk axis (SW–NE), the oblique conical equal-area projection and the oblique azimuthal equal-area projection were expected to offer the best results. Distortion at the extremities proved to be marginally better for an azimuthal equal-area projection with centre at 48°N, 9°E, close to the small town of Tuttlingen in Bavaria, than for the conical equal-area projection that Maling considered in his analysis, with the meta-pole at 55°N, 43°E, and the standard lines midway between the middle and limiting meta-parallels (Bornholm in the NE and Cape St-Vincent in the SW). Maling therefore proposed the azimuthal equal-area projection centred on Tuttlingen as the best choice for the mapping of the EU, with a maximum scale error of 1.31%, and a maximum angular distortion of 1°.5 at the extremities of the area (figure 5.38).

Meanwhile, with the reunion of Germany in 1990, and the enlargement of the EU with three new member states (Austria, Finland and Sweden) in 1993, the borders of the EU have changed considerably, and the distortion pattern of Maling's azimuthal equal-area projection no longer matches the shape of the area. The northern extremities of the Scandinavian countries have a maximum scale error and a maximum angular distortion that are twice as high as for the north-west, south-west and south-east extremities of the territory. As such, the problem of selecting a suitable map projection for the EU should be reconsidered.

Standard equal-area projections

In most atlases and geography textbooks, maps of Europe are drawn on Albers' conical equal-area projection, Bonne's pseudoconical equal-area projection, or the oblique version of Lambert's azimuthal equal-area projection. The use of the conical equal-area projection follows from a well-known, very simple rule in cartographic design, which states that conical projections are the best choice for the mapping of areas in the middle latitudes (Maling 1992, p. 229, see also section 6.1). However, the latitudinal extent of Europe is far too wide for the conical projection to avoid extreme distortion in those areas that are far away from the two standard parallels (figure 5.39). The use of Bonne's pseudoconical projection raises similar problems. While Bonne's projection is free of distortion along the central meridian and along one chosen standard parallel, distortion rapidly increases away from either line, leading to relatively high distortion values in the north-west, south-west, and south-east corners of the EU (figure 5.40).

Of the three equal-area projections that are mostly used for maps of Europe, the oblique azimuthal equal-area projection proves to have the least extreme distortion, yet the variation in scale along the boundaries of the EU is still quite high (figure 5.41). Table 5.24 lists distortion values for a set of extreme points for the optimised version of all three projections. As before, optimisation was accomplished by minimisation of the mean finite scale distortion for a set of 5000 randomly selected distances covering the entire area.

Less familiar aspects of standard equal-area projections

While oblique azimuthal projections are frequently found in contemporary atlases, oblique aspects of conical or pseudoconical projections are rarely used. This may be a reflectance of strong traditions in map making or, expressing a more pessimistic view, it may also point at a general lack of interest in the problems of map projection choice among present map makers. Because azimuthal projections have a radial distortion pattern, it is fairly easy to choose an appropriate location for the centre once the spatial extent of the area to be mapped is known. This explains why oblique azimuthal projections have been used since long. For conical and pseudoconical projections, it is not that simple to choose oblique aspect parameter values that are optimally adapted to the region to be mapped, without using some numerical technique to reduce distortion. Hence, it is not surprising that these oblique aspects were not frequently used in traditional cartographic practice. Today, however, there is really no reason why optimised aspects of conical and pseudoconical projections should not be used for the small-scale mapping of particular regions of the globe. As mentioned before, a simple move of the distortion pattern of the projection with respect to the Earth's surface may be a very efficient way of reducing distortion within the area to be mapped.

To see what can be gained from aspect optimisation for the mapping of the EU, optimal aspects of the conical equal-area projection and the Bonne projection

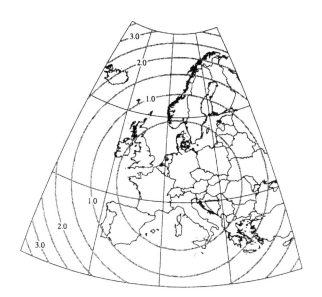

Figure 5.38 Maling's azimuthal equal-area projection for the European Union (with origin at 48°N, 9°E), with lines of constant maximum scale error (%).

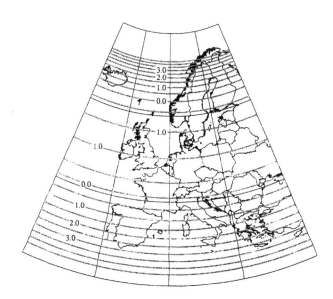

Figure 5.39 Conical equal-area projection with optimised parameters for the European Union (standard latitudes 46°N, 62°N), with lines of constant maximum scale error (%).

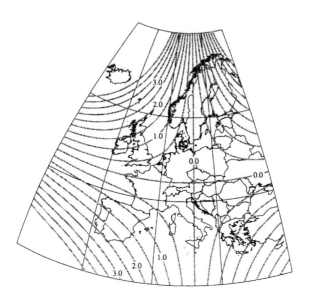

Figure 5.40 Bonne's pseudoconical equal-area projection with optimised parameters for the European Union (standard latitude 48°N, central meridian 14°E), with lines of constant maximum scale error (%).

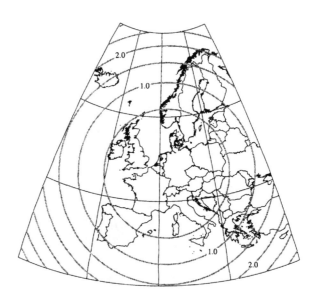

Figure 5.41 Oblique azimuthal equal-area projection with optimised parameters for the European Union (origin at 50°N, 7°E), with lines of constant maximum scale error (%).

have been derived by minimisation of the mean distortion of finite distance (figure 5.42). Table 5.25 lists the mean finite scale error for all distances (%), and the maximum local scale error (%) within the boundaries of the area, for the optimised oblique versions of the azimuthal, the conical and the Bonne projection. For comparison, the results for the normal conical and the normal Bonne projection, both with optimised parameters, have also been included. As can be seen, the use of oblique aspects may lead to a substantial reduction of overall distortion. Especially the oblique Bonne has a very low mean finite distortion. This is confirmed by the shaping of the isocols, which are well adapted to the general outline of the EU (figure 5.42b). Extreme local scale error, however, is still 1.8%, which is almost as high as for the optimised azimuthal equal-area projection. The meta-pole of the oblique Bonne is situated at 52°N, 29°E, and introduces a small discontinuity in the map. Although the gap is not located in the area covered by the EU, its presence may be considered as inconvenient for small-scale mapping.

Table 5.24 Maximum scale error and maximum angular distortion at the extremities of the EU for low-error versions of Albers' conical equal-area projection, Bonne's pseudoconical equal-area projection, and the oblique azimuthal equal-area projection.

		Maximum scale error (%)	Maximum angular distortion (°)
Albers' conical equal-area	Utsjoki (Finland)	4.83	5.40
	Thorshavn (Faeroer)	0.13	0.15
	Stornoway (Hebrides)	0.90	1.03
	Sagres (Portugal)	2.49	2.82
	Tarifa (Spain)	2.86	3.23
	Ragusa (Sicily)	2.52	2.86
	Sitia (Crete)	3.16	3.57
	Rodos (Rhodes)	2.71	3.06
Bonne's pseudoconical equal-area	Utsjoki (Finland)	3.06	3.45
	Thorshavn (Faeroer)	3.03	3.42
	Stornoway (Hebrides)	2.11	2.39
	Sagres (Portugal)	2.70	3.05
	Tarifa (Spain)	2.52	2.85
	Ragusa (Sicily)	0.06	0.07
	Sitia (Crete)	1.60	1.82
	Rodos (Rhodes)	1.71	1.95
Oblique azimuthal equal-area	Utsjoki (Finland)	1.87	2.12
	Thorshavn (Faeroer)	0.77	0.88
	Stornoway (Hebrides)	0.49	0.56
	Sagres (Portugal)	1.15	1.31
	Tarifa (Spain)	1.07	1.22
	Ragusa (Sicily)	0.77	0.88
	Sitia (Crete)	1.59	1.80
	Rodos (Rhodes)	1.61	1.83

Table 5.25 Mean distortion of finite distance and maximum scale error at the extremities of the EU for low-error versions of various standard equal-area projections.

	Mean distortion of finite distance (%)	Maximum scale error (%)
Oblique azimuthal	0.26	1.87
Oblique conical	0.24	5.94
Oblique Bonne	0.17	1.80
Normal conical	0.55	4.83
Normal Bonne	0.44	3.06

As in the example of Eurafrica, which has been discussed above, it is also clear from this case that the minimisation of the mean distortion of finite distances not necessarily leads to a lowering of extreme scale error within the area to be mapped. The oblique conical projection has a higher extreme scale error than the optimal version of this projection in the normal aspect, although the mean finite scale distortion is reduced by more than 50%. This again demonstrates that optimised graticules should not be evaluated solely on the basis of their mean distortion values, but also by examination of distortion at the extremities of the area. The same holds for the optimal equal-area transformations that will be discussed below.

A close look at a map of Europe shows that the EU has rather irregular boundaries and lacks compactness. The shape of the area more or less approximates a spherical triangle that is slightly inclined with respect to the direction of the meridians. One therefore might expect to obtain less overall distortion, and a better adaptation of the isocols to the general outline of the EU, by applying polynomial equal-area transformations of the same type as those that were used for the mapping of Eurafrica, and for maps of the entire continental area in the previous parts of this chapter. Starting with the map coordinates for the optimised version of the oblique azimuthal equal-area projection of the EU with centre at 50°N, 7°E, again various optimal equal-area transformations, as defined by equations (4.84) and (4.88), can be derived by working with polynomials of different order, and by changing the symmetry conditions of the transformation.

Instead of directly applying the double polynomial transformation to the coordinates of the oblique azimuthal projection, as it was done before, this time the x- and y-axes of the projection will be rotated about an angle θ in the map plane, before the transformation is carried out. This additional rotation increases the flexibility of the transformation, and is especially important for symmetric transformations since it allows the symmetry axes of the distortion pattern to become aligned with the directions of maximum and/or minimum extent of the area to be mapped. Both the angle of rotation, and the polynomial coefficients that define the double transformation, can be optimised simultaneously. For the EU in particular, the two axes of maximum extent do not coincide with the north and east direction, as for Eurafrica, but are more or less aligned with the NW–SE and NE–SW directions. Hence, the optimal angle between the y-axis of the rotated

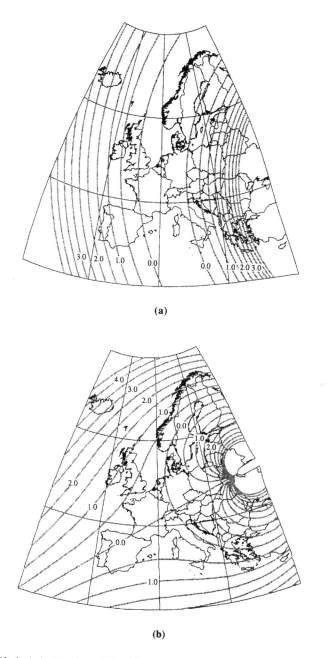

(a)

(b)

Figure 5.42 Optimised versions of the oblique conical equal-area projection (a) and the oblique Bonne projection (b) for the European Union, with lines of constant maximum scale error (%).

coordinate system and the north direction is expected to be substantially different from zero, which justifies the use of the rotation.

Optimisation of the affine coefficient for a simple first-order transformation, defined by equations (4.89), combined with a rotation of the coordinate axes, produces a graticule with oval isocols, with their major axes making an angle of 30° with the central meridian (figure 5.43). Since the EU has two, almost perpendicular directions of maximum extent, the distortion pattern does not depart much from the original circular shape. The mean distortion for the first-order transformation is only slightly lower than for the oblique azimuthal projection (table 5.26). Minimisation of distortion for a second-order and third-order transformation with symmetry about the rotated y-axis yields graticules with lower mean finite distortion. The distortion patterns are more closely adapted to the general outline of the EU, which leads to a better distribution of scale error (figures 5.44 and 5.45). Again, however, it should be emphasised that extreme scale error along the boundaries of the area is not necessarily lowered by an increase of the order of the polynomials. Although the mean finite scale error for the third-order transformation is much lower than for the original oblique azimuthal projection, the extreme scale error is higher (compare figure 5.41 with figure 5.45).

Applying the polynomial transformation without imposing symmetry about one of the coordinate axes increases the possibilities for a better adaptation of the distortion pattern to the shape of the area. Figures 5.46 and 5.47 show the results obtained for a second-order and third-order transformation. As can be seen, maximum scale error in the south-east of the EU (Greece) is lower than for a symmetric arrangement of the isocols. Distortion in the north-west (Great Britain

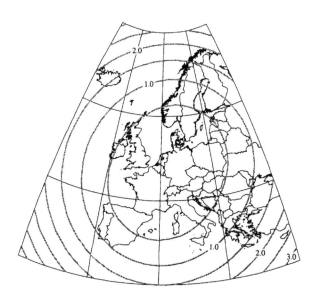

Figure 5.43 First-order low-error transformation of the oblique azimuthal equal-area projection for the European Union, with lines of constant maximum scale error (%).

Figure 5.44 Second-order one-fold symmetric low-error transformation of the oblique azimuthal equal-area projection for the European Union, with lines of constant maximum scale error (%).

Figure 5.45 Third-order one-fold symmetric low-error transformation of the oblique azimuthal equal-area projection for the European Union, with lines of constant maximum scale error (%).

and Ireland) is somewhat increased. The result obtained with the second-order transformation is particularly interesting for its favourable distribution of scale error. Although the graticule has a mean distortion of finite scale that is only slightly lower than for the oblique azimuthal projection (0.22%), the extreme scale error along the outer boundaries is only 1.21%. This implies that the maximum angular distortion does not exceed 1.4° throughout the whole of the EU. All the isocols have a triangle-like shape that matches the general extent of the area. The isocol of 1% scale error passes nicely through the extremities in the north, the south-west and the south-east of the map. As a result, the variation in maximum scale error along the edges of the area is very small (table 5.27).

For the third-order transformation, the mean scale error is much lower than for the second-order result (0.15% against 0.22%), yet the variation in maximum scale error for the extreme points is higher (table 5.27). This is also clear from the pattern of the isocols, which is strongly elongated along the major axis of the area (figure 5.47). Again this indicates that the optimisation of polynomial transformations by minimisation of finite distance distortion will not, in general, produce a pattern of isocols that follows the outline of the area, at least not for areas with complex boundary definitions. Using higher-order polynomials will create graticules with a more complex pattern of isocols, which will guarantee a further reduction of overall distortion. However, one cannot expect the variation of scale error along the boundaries of the area to be a minimum. To demonstrate this, also a fifth-order transformation has been applied (figure 5.48). As can be seen, the pattern of isocols is more closely adapted to the shape of the area than for a third-order transformation. With the exception of the Faeroer Islands, which are located

Figure 5.46 Second-order low-error transformation of the oblique azimuthal equal-area projection for the European Union, with lines of constant maximum scale error (%).

Figure 5.47 Third-order low-error transformation of the oblique azimuthal equal-area projection for the European Union, with lines of constant maximum scale error (%)

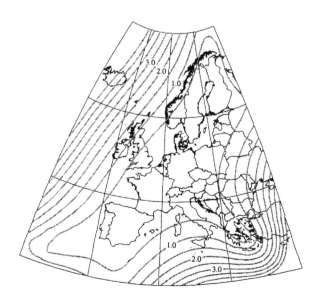

Figure 5.48 Fifth-order low-error transformation of the oblique azimuthal equal-area projection for the European Union, with lines of constant maximum scale error (%).

very excentrically, the whole area has less than 1.5% scale error, while the mean error of finite distances is less than half the value obtained for the original oblique azimuthal projection (0.12% instead of 0.26%). Nevertheless the range of scale error along the boundaries remains high, with a minimum of 0.10% for Utsjoki (Finland), and a maximum of 2.05% for Thorshavn (Faeroer) (table 5.27).

Simplifying the boundary definition

The spatial distribution of scale error in the optimised graticules is, of course, strongly related to the way in which distances for which distortion is minimised are selected. In the above examples, starting and end points for each distance are located within the area of the EU. Since most of the area stretches along the SW–NE axis, it is not surprising that the distortion pattern of the optimised graticules aligns with this direction, and produces the highest scale error in areas that are far away from this line. Distortion is extreme for Faeroer and for the Greek islands, which are geographically isolated from the continental EU, and have little impact on the minimisation process. One might expect extreme distortion in these far away areas to be reduced by selecting starting and end points for the distances that are used for error minimisation within a smooth polygon that encloses the entire area of the EU. In that case no isolated areas will occur, and error is likely to become more evenly distributed.

To evaluate the impact of using a single boundary definition, a polygon was constructed that closely follows the general outline of the EU (figure 5.49). Minimisation of the mean distortion of finite distance for a second-, third- and fifth-order equal-area transformation, with distances selected randomly within the boundary of this polygon, produces the graticules shown in figures 5.50–5.52. Parameter values and mean distortion of finite distance for the three transformations are listed in table 5.28. Starting point for the optimisation again was the optimal aspect of the oblique azimuthal equal-area projection for the area, this time with centre at 49°N, 7°E. As can be seen, the result for the second-order transformation does not differ much from the result that was obtained before (figure 5.46), except that the distortion pattern is somewhat shifted to the west, lowering the distortion values for Great Britain, Ireland and Faeroer at the expense of the Greek area (for Rodos the maximum scale error is 1.38%). Also for the northern countries of the EU scale distortion is increased. Much better solutions are obtained with the third-order and fifth-order transformation. Compared to the previous results (figure 5.47, figure 5.48), the pattern of isocols better follows the general outline of the EU, as it is defined by the polygon in figure 5.49, leading to less extreme scale distortion along the boundary of the area. Except from Rodos, where the scale error reaches 1.08% for the third-order and 1.13% for the fifth-order transformation, scale distortion is well below 1% for the entire EU (table 5.29).

From tables 5.28 and 5.29 it is also clear that no real improvement is obtained with the fifth-order transformation, neither in terms of the mean distortion of finite distance, nor in terms of maximum scale error for extreme points. Comparing the distortion patterns obtained with the third-order and fifth-order transformation (figures 5.51 and 5.52) also shows that both produce very similar results. Although no general conclusions can be drawn from this, it surely

Table 5.26 Parameter values and mean distortion of finite distance for various low-error equal-area transformations of the oblique azimuthal equal-area projection for the EU.

	Parameter values		Mean distortion of finite distance (%)
FIRST-ORDER TRANSFORMATION (*)	$C_1 = 0.9991$	$C'_1 = 1.0009$ $\theta = 30.48°$	0.25
SECOND-ORDER TRANSFORMATION (*)			
with one-fold symmetry	$C_1 = 1.0151$	$C'_1 = 1.0164$ $C'_2 = 0.0086$ $\theta = 17.23°$	0.23
without symmetry	$C_1 = 0.9997$ $C_2 = 0.0095$	$C'_1 = 1.0011$ $C'_2 = 0.0056$ $\theta = 29.27°$	0.22
THIRD-ORDER TRANSFORMATION (*)			
with one-fold symmetry	$C_1 = 1.0004$ $C_3 = -0.0170$	$C'_1 = 1.0001$ $C'_2 = 0.0025$ $C'_3 = 0.0360$ $\theta = 26.69°$	0.17
without symmetry	$C_1 = 1.0049$ $C_2 = 0.0085$ $C_3 = -0.0198$	$C'_1 = 1.0046$ $C'_2 = 0.0010$ $C'_3 = 0.0368$ $\theta = 30.61°$	0.15
FIFTH-ORDER TRANSFORMATION (*)			
with one-fold symmetry	$C_1 = 1.0016$ $C_3 = -0.0477$ $C_5 = 0.3216$	$C'_1 = 1.0007$ $C'_2 = -0.0030$ $C'_3 = 0.0622$ $C'_4 = 0.0701$ $C'_5 = -0.2578$ $\theta = 26.69°$	0.16
without symmetry	$C_1 = 1.0059$ $C_2 = 0.0018$ $C_3 = -0.0595$ $C_4 = 0.1665$ $C_5 = 0.2369$	$C'_1 = 1.0052$ $C'_2 = 0.0000$ $C'_3 = 0.0519$ $C'_4 = -0.0035$ $C'_5 = -0.1039$ $\theta = 37.77°$	0.12

(*) Parent projection: oblique azimuthal equal-area projection with meta-pole at 50°N, 7°E

indicates that the most spectacular improvements are likely to be obtained with low-order polynomials. Including higher-order terms in the transformation may lead to a further reduction of the mean scale distortion for the area to be mapped. This is

Table 5.27 Maximum scale-error and maximum angular distortion at the extremities of the EU for the second-order, third-order, and fifth-order low-error equal-area transformation of the oblique azimuthal equal-area projection (no symmetry conditions imposed).

		Maximum scale error (%)	Maximum angular distortion (°)
SECOND-ORDER	Utsjoki (Finland)	1.10	1.25
	Thorshavn (Faeroer)	1.21	1.38
	Stornoway (Hebrides)	1.00	1.14
	Sagres (Portugal)	0.99	1.13
	Tarifa (Spain)	1.12	1.27
	Ragusa (Sicily)	0.75	0.86
	Sitia (Crete)	0.99	1.13
	Rodos (Rhodes)	1.04	1.18
THIRD-ORDER	Utsjoki (Finland)	0.59	0.67
	Thorshavn (Faeroer)	1.36	1.55
	Stornoway (Hebrides)	0.98	1.12
	Sagres (Portugal)	0.50	0.57
	Tarifa (Spain)	0.16	0.19
	Ragusa (Sicily)	1.01	1.15
	Sitia (Crete)	1.74	1.97
	Rodos (Rhodes)	1.73	1.97
FIFTH-ORDER	Utsjoki (Finland)	0.10	0.12
	Thorshavn (Faeroer)	2.05	2.32
	Stornoway (Hebrides)	1.44	1.64
	Sagres (Portugal)	0.13	0.14
	Tarifa (Spain)	0.18	0.21
	Ragusa (Sicily)	1.42	1.61
	Sitia (Crete)	1.14	1.30
	Rodos (Rhodes)	1.39	1.59

especially true if boundary conditions are complex, as has been shown in the previous part of this section, where distances for error minimisation were defined with their starting and end points within the irregular boundaries of the member states. One should however bear in mind that an increase of the flexibility of the transformation may result in higher distortion values for some extreme points, which puts a practical limit to the use of higher-order terms.

5.5 SOME GENERAL REMARKS ON MAP PROJECTION OPTIMISATION

When selecting a map projection for a particular purpose, there are two important

Figure 5.49 Smooth polygon defining the outline of the European Union.

Figure 5.50 Second-order low-error transformation of the oblique azimuthal equal-area projection for the European Union, with lines of constant maximum scale error (%) (single boundary definition).

Figure 5.51 Third-order low-error transformation of the oblique azimuthal equal-area projection for the European Union, with lines of constant maximum scale error (%) (single boundary definition).

Figure 5.52 Fifth-order low-error transformation of the oblique azimuthal equal-area projection for the European Union, with lines of constant maximum scale error (%) (single boundary definition).

Table 5.28 Parameter values and mean distortion of finite distance for the second-order, third-order, and fifth-order low-error equal-area transformation of the oblique azimuthal equal-area projection for the EU (single boundary definition).

	Parameter values		Mean distortion of finite distance (%)
SECOND-ORDER TRANSFORMATION (*)			
without symmetry	$C_1 = 1.0004$	$C'_1 = 1.0010$	0.29
	$C_2 = -0.0053$	$C'_2 = 0.0070$	
		$\theta = 2.21°$	
THIRD-ORDER TRANSFORMATION (*)			
without symmetry	$C_1 = 0.9937$	$C'_1 = 0.9939$	0.22
	$C_2 = -0.0066$	$C'_2 = 0.0043$	
	$C_3 = -0.0440$	$C'_3 = -0.0137$	
		$\theta = -9.70°$	
FIFTH-ORDER TRANSFORMATION (*)			
without symmetry	$C_1 = 1.0011$	$C'_1 = 1.0012$	0.22
	$C_2 = -0.0053$	$C'_2 = 0.0028$	
	$C_3 = -0.0444$	$C'_3 = -0.0135$	
	$C_4 = -0.0217$	$C'_4 = 0.0175$	
	$C_5 = 0.0306$	$C'_5 = 0.0036$	
		$\theta = -9.09$	

(*) Parent projection: oblique azimuthal equal-area projection with meta-pole at 49°N, 7°E
(mean distortion of finite distance for the area bounded by the smooth polygon is 0.32%)

elements that should be taken into account, i.e. map use and map distortion. While map projection is mostly concerned with reducing distortion, one must make sure that the projection has all the properties that are considered essential for the application one has in mind. As stated by Bugayevskiy and Snyder (1995, p. 193), the best projections are therefore "... those satisfying in an optimum way an entire group of requirements for projections in accordance with the particular purpose of the map being designed...". In practice, the selection of a suitable projection for a small-scale map should start with the formulation of a number of requirements, which may relate to the geometry of the graticule (shape of the meridians and the parallels, representation of the pole, aspect of the projection, ...), as well as to its distortion properties (equal-area, conformal, equidistant along one or more lines, ...). Once a set of requirements has been defined, the spectrum of possible map projections will already be considerably narrowed. Quantitative analysis, based on the calculation of local or global distortion values, may then help to select the most suitable projection.

When performing distortion analysis, certain requirements may be specified. These may be simple non-quantitative requirements, such as the demand for minimum visible distortion. On the other hand, one may also specify certain

Table 5.29 Maximum scale-error and maximum angular distortion at the extremities of the EU for the second-order, third-order, and fifth-order low-error equal-area transformation of the oblique azimuthal equal-area projection (single boundary definition).

		Maximum scale error (%)	Maximum angular distortion (°)
SECOND-ORDER	Utsjoki (Finland)	1.37	1.56
	Thorshavn (Faeroer)	0.96	1.09
	Stornoway (Hebrides)	0.72	0.83
	Sagres (Portugal)	1.06	1.21
	Tarifa (Spain)	1.15	1.31
	Ragusa (Sicily)	0.59	0.67
	Sitia (Crete)	1.25	1.42
	Rodos (Rhodes)	1.38	1.58
THIRD-ORDER	Utsjoki (Finland)	0.79	0.91
	Thorshavn (Faeroer)	0.88	1.00
	Stornoway (Hebrides)	0.45	0.51
	Sagres (Portugal)	0.85	0.97
	Tarifa (Spain)	0.37	0.42
	Ragusa (Sicily)	0.77	0.88
	Sitia (Crete)	0.47	0.53
	Rodos (Rhodes)	1.08	1.23
FIFTH-ORDER	Utsjoki (Finland)	0.76	0.87
	Thorshavn (Faeroer)	0.82	0.94
	Stornoway (Hebrides)	0.45	0.51
	Sagres (Portugal)	0.80	0.92
	Tarifa (Spain)	0.44	0.50
	Ragusa (Sicily)	0.77	0.88
	Sitia (Crete)	0.48	0.54
	Rodos (Rhodes)	1.13	1.29

limiting values of distortion that should be satisfied on the map. In both cases, numerical optimisation, based on some global distortion measure, may be very helpful to decide which of the remaining projections is best. Optimisation of map projection parameters for different candidate projections allows an objective comparison of the suitability of each projection for the mapping of a particular region. If none of the candidate projections satisfies the distortion requirements that have been imposed (e.g. too much visible distortion, too much scale variation within the area), different map projection modifications or transformations may be applied, from the definition of two standard lines, to the use of higher-order polynomial transformations with well-defined characteristics. Especially for the latter numerical optimisation is a prerequisite.

Although map projection optimisation should be an essential part of any procedure for map projection selection, one must realise that the minimisation of a global distortion measure will not necessarily produce a useful map, or a map that serves its purpose best. Several examples, some of which have occurred in this chapter, can be listed. When optimising the aspect of a projection for a world map in an attempt to reduce overall distortion, the outer meta-meridian of the graticule may intersect one of the major continents, and cause a split of the area in two parts that are located on opposite sides of the graticule (figure 5.24). This is usually not considered an optimal solution, and a small adjustment of the centre of the projection will be necessary. Another example concerns the use of conic type projections for continental or regional maps. Since both azimuthal and cylindrical map projections are limiting cases of the conical projection, the latter will always produce the least distortion for a given area. This was demonstrated for North and South America (see section 5.4.1). One might be tempted to think that this automatically eliminates the problem of having to make a choice among these three types of projections. However, optimisation of the conical projection will not always produce acceptable results. As the outline of the area to be mapped becomes more compact, the meta-pole of the conical projection very often will be located within the area itself. This will cause a wedge-shaped interruption of the map where the cone opens (Snyder, 1985, p. 103). In such case one will be forced to use one of the other two projections.

Other problems with optimisation are related to the optimisation criterion that is used. Optimisation of the parameters of a map projection based on the minimisation of a mean overall distortion measure will, in general, not yield the lowest possible range of scale factors, especially not if the outline of the area is highly irregular, or asymmetrical with respect to the distortion pattern of the projection. A good example of this is the South-American continent. When mapped on a projection with a straight central meridian, as is very often done, optimisation of the parameters will always yield a "central" longitude that is clearly off-centre (too close to the western boundary of the continent). It is obvious that the range of scale factors can be substantially reduced by moving the central meridian somewhat more to the east, this however at the cost of a higher overall distortion. Problems of this kind may be less pronounced if the symmetry of the graticule is renounced and polynomial transformation is applied, since in that case the isocols can be reshaped to follow the outline of the area more closely.

For global mapping, the shortcomings of using mean overall distortion measures have been shown earlier in this chapter. Due to the irregular distribution of the continents, minimisation of overall distortion may excessively deform those areas that are located far from the centre of gravity of the landmasses. When using polynomial transformation, parts of the map outside the continental area may even be folded. To produce maps with a more balanced pattern of distortion, one may have to impose additional constraints, to control the process of optimisation. This makes it clear that the use of optimisation techniques in a procedure for map projection selection should always be linked to a verification process, which may lead to a re-definition of constraints. The process of verification and adjustment of constraints may have to be repeated several times before an optimal solution is reached. If no satisfactory solution can be found, one may even have to change some of the initial conditions that have been imposed by the purpose of the map, in order to produce an acceptable compromise. In the next chapter, which deals with map projection selection, we will further enlarge upon this issue.

CHAPTER SIX

Semi-automated
map projection selection

The selection of a suitable map projection is an important issue in small-scale map design, especially for the mapping of large areas, where map projection distortion exceeds the threshold of visibility. Over the years, a great number of map projections have been proposed. Even for the skilled cartographer, who has a good knowledge of map projection principles, choosing among this variety of existing projections is not an easy task. Map projection selection interferes with several other variables in map design. As such it is not possible to compile a magic table telling us unambiguously which projection is best for a given application. With the increasing use of geographical information systems, and the development of new mapping software for personal computers, cartographic tools are coming within the reach of an ever-growing group of users that are unfamiliar with the principles of cartographic design. This has led some cartographers to determine what is currently known about map design from years of theoretical and practical research, and try to translate this knowledge into a set of rules that can be built into microcomputer-based software. Most efforts in this direction have concentrated on cartographic generalisation and the automated placement of text features (Buttenfield and McMaster, 1991; Freeman, 1991). Some researchers have also addressed the problem of map projection selection, taking into account the limitations imposed by the digital environment, as well as the new opportunities for map projection development that are offered by the computer. These efforts will be discussed in more detail in the first part of this chapter, together with more traditional approaches that have been suggested in the past.

The choice of a map projection is strongly related to the purpose of the map. The area that will be covered, the ways in which the map is going to be used, and the intended audience for whom the map is targeted are all major elements in the decision process. Hence selection should start with the definition of a unique set of requirements (map projection properties) that best suits the purpose of the mapping (see also section 5.5). As will be shown, most map projection selection schemes that have been proposed so far are either too rigid in the definition of their rules, or include only part of the map projection properties that might be considered useful. Also many of these schemes do not lead the user to a unique solution, because of the large number of existing map projections from which a choice has to be made. Using these schemes, one often ends up with various candidate projections that may be very alike, with no simple means of deciding which one is to be preferred. In this chapter, a new selection scheme will be presented that is based on the most important map projection features that may be required. The proposed strategy is founded on the combination of feature-based selection (to define the map projection) and numerical optimisation (to reduce overall distortion), and makes use of the techniques that have been described and applied in the two previous chapters.

6.1 DIFFERENT APPROACHES TO MAP PROJECTION SELECTION

Any attempt to define a framework for map projection selection automatically entails a grouping of map projections that is based on the criteria that are used in the selection process. Hence, there is a strong relationship between map projection selection and map projection classification. As stated by Nyerges and Jankowski (1989), selection is perhaps the ultimate reason for devising classifications. Already in the first chapter of this study, in summarising the requirements for useful classification schemes, it has been emphasized that classification should facilitate selection. However, map projection selection is not the reverse of classification. While in a classification scheme various projections may have the same properties, map projection selection must lead to a unique solution. Moreover, map projection selection involves trade-offs that cannot be expressed in a formal way. This complicates any systematic treatment of the subject, and makes the selection process somewhat difficult to comprehend. In spite of the practical importance of map projection selection, only very few authors have treated the subject in any detail. Knowledge on map projection selection mostly appears as a heterogeneous and partly inconsistent collection of rules, repeatedly found in general textbooks on cartography or map projection. Many authors only provide a summary of the map projections that are most frequently used for the mapping of various areas and for different map purposes. These listings are of limited use to a novice cartographer, since they do not provide any insight in the reasoning that brings the cartographer to choose for a particular projection.

One of the most elementary methods for map projection selection, which is found in each textbook on cartography, immediately derives from the three-fold geometric classification of map projections into an azimuthal, a conical and a cylindrical class (see section 1.4.1), and is based on the following three rules (Maling, 1992, p. 229):

1. Azimuthal projections should be used for maps of the polar regions.
2. Conical projections are to be preferred for areas in temperate latitudes.
3. Cylindrical projections are best for the mapping of tropical regions.

These simple rules follow from the position of the point or line of zero distortion in the normal, unmodified version of each of the three projections. According to Maling, the principles have been applied to the design of sheet and atlas maps since the sixteenth century. He refers to them as "... one of the classical foundations of cartographic design". Of course, these rules ignore the existence of many other types of projections that can be used. They also do not take account of the fact that map projection distortion can be substantially reduced through a change of aspect of the projection (see section 2.6). If transverse and oblique aspects are also taken into consideration, then the projection class can be chosen in terms of the shape of the region, regardless of its geographical location. A modern, more flexible interpretation of the above rules, no longer limited by the aspect, could be stated as follows:

1. Azimuthal projections are best for the representation of areas with a circular outline.
2. Conical projections are to be preferred for areas that extend along a small circle.
3. Cylindrical projections are most suited for areas that extend along a great circle.

Young (1920) suggested a simple measure, based on the comparison of the maximum and minimum extent of the area to be mapped, to facilitate the choice between an azimuthal projection, and one of the other two projection types. Nowadays, it is better to rely on techniques for aspect optimisation to find out which of the three projection types is the best, in case there should be any doubt (see section 5.4.1).

While the geometric characteristics of the region to be mapped will determine the choice of class and aspect, different elements related to the function, and the intended use of the map (content, symbolisation, audience, ...), will specify the properties the map projection should have. Although many useful properties can be listed (see section 6.2), most authors only refer to special distortion properties, distinguishing between conformal, equal-area, and equidistant projections. As was demonstrated in section 2.4.2, conformal projections show an extreme distortion of area, while equal-area projections excessively distort angular relationships. As such, both types of projections can be considered as being the two limits of choice. In large-scale mapping the use of conformal projections is universally accepted. Equal-area projections are often used for the small-scale mapping of statistical data. Between these two extremes there are a variety of other map projections that do not have the extreme distortions which are characteristic of conformal and equal-area maps.

Equidistant projections prove to occupy a central position within the continuum of all map projections (see also section 2.5.1). Hence, they provide a useful compromise for maps that do not necessarily have to be conformal or equal-area. Putting equidistant projections halfway between the conformal projections and the equal-area projections, Maling (1992, p. 238) further distinguishes projections with small angular distortion, and projections with small exaggeration of area, and associates different map uses with the five obtained groups. As shown in table 6.1, however, no definite rules exist to match a certain type of map with one of the five groups of projections. Detailed study of the distortion requirements for various types of small-scale maps indeed shows that it is difficult to make a proper choice, not in the least because small-scale maps very often have to serve more than one purpose at the same time (Bugayevskiy and Snyder, 1995, p. 240).

Snyder (1987a, p. 34) is one of the few authors who treat the problem of map projection selection in a systematic way. To facilitate the selection process, he presents a decision tree that is based on: (a) size, shape, orientation and location of the region to be mapped, (b) special distortion properties (conformal, equal-area,

Table 6.1 Special distortion properties and main uses of map projections (Maling, 1992).

MAIN USES	SPECIAL PROPERTY	Conformal projections	(Projections with small angular distortion)	Equidistant projections	(Projections with small exaggeration of area)	Equal-area projections
Navigational charts		x				
Topographical, military and large-scale maps		x				
Synoptic meteorological charts		x				
Small-scale strategic planning maps		x	x	x		
Climatic and oceanographic distribution maps		x	x	x		
General reference maps			x	x	x	x
Atlas maps				x	x	
Statistical distribution maps						x

equidistant, correct scale along a chosen great-circle), and (c) application-specific considerations (e.g. straight rhumb lines, straight great-circle routes, interrupted designs) (table 6.2). Snyder recognizes that world maps cannot be satisfactorily represented by means of conic type projections, and that other selection criteria may be involved in global mapping than in continental or regional mapping (e.g. the decision to interrupt the graticule or not). He therefore makes a clear distinction between world maps and maps of smaller areas. For world maps many types of projections are listed, depending on the application. For smaller areas he recommends the use of conic type projections. Maps of a hemisphere, which usually have a circular outline, are also treated as a separate class. For these maps, Snyder recommends the use of azimuthal projections.

While table 6.2 may be useful for the practitioner, it was not Snyder's endeavour to present an all-inclusive scheme for map projection selection. Snyder started with a set of commonly used map projections, and developed a practical scheme to indicate in which cases each projection should best be used. This makes that some interesting combinations of properties are not included, since none of the projections that are listed satisfies them. For example, of the rather large group of world map projections that are neither equal-area, nor conformal, only Miller's cylindrical and Robinson's projection are listed. Since both projections have a pole line they are less suited for the derivation of oblique aspects (see section 2.6). Other projections of the same group (e.g. Aitoff's or Apianus' projection, which

Table 6.2 Snyder's decision tree for map projection selection (Snyder, 1987a).

WORLD
    A. Conformal (gross area distortion)
        (1) Constant scale along Equator
            Mercator
        (2) Constant scale along meridian
            Transverse Mercator
        (3) Constant scale along oblique great circle
            Oblique Mercator
        (4) Entire Earth shown
            Lagrange
            August
            Eisenlohr
    B. Equal-area
        (1) Standard without interruption
            Hammer
            Mollweide
            Eckert IV or VI
            McBryde or McBryde–Thomas variations
            Boggs Eumorphic
            Sinusoidal
            misc. pseudocylindricals
        (2) Interrupted for land or ocean
            any of above except Hammer
            Goode Homolosine
        (3) Oblique aspect to group continents
            Briesemeister
            Oblique Mollweide
    C. Equidistant
        (1) Centred on pole
            Polar Azimuthal Equidistant
        (2) Centred on a city
            Oblique Azimuthal Equidistant
    D. Straight rhumb lines
        Mercator
    E. Compromise distortion
        Miller Cylindrical
        Robinson

both represent the world within an ellipse) are more apt to that purpose. Snyder, however, does not include oblique aspects of non-equal-area world map projections for the simple reason that these are not commonly applied. Another, example which indicates that Snyder restricted the scheme to more traditional choices, is the absence of the oblique conical map projection as a worthy alternative for the more frequently used oblique cylindrical projection (see also section 5.4.1). The number of map projection properties included in Snyder's scheme is limited. Several useful

Table 6.2 continued

HEMISPHERE
    A. Conformal
        Stereographic (any aspect)
    B. Equal-area
        Lambert Azimuthal Equal-Area (any aspect)
    C. Equidistant
        Azimuthal Equidistant (any aspect)
    D. Global look
        Orthographic (any aspect)

CONTINENT, OCEAN OR SMALLER REGION
    A. Predominant east–west extent
        (1) Along Equator
            Conformal: Mercator
            Equal-Area: Cylindrical Equal-Area
        (2) Away from Equator
            Conformal: Lambert Conformal Conic
            Equal-Area: Albers Equal-Area Conic
    B. Predominant north–south extent
        Conformal: Transverse Mercator
        Equal-Area: Transverse Cylindrical Equal-Area
    C. Predominant oblique extent (for example: North America, South America,
       Atlantic Ocean)
        Conformal: Oblique Mercator
        Equal-Area: Oblique Cylindrical Equal-Area
    D. Equal extent in all directions (for example: Europe, Africa, Asia, Australia,
       Antarctica, Pacific Ocean, Indian Ocean, Arctic Ocean, Antarctic Ocean)
        (1) Centre at pole
            Conformal: Polar Stereographic
            Equal-Area: Polar Lambert Azimuthal Equal-Area
        (2) Centre along Equator
            Conformal: Equatorial Stereographic
            Equal-Area: Equatorial Lambert Azimuthal Equal-Area
        (3) Centre away from pole or Equator
            Conformal: Oblique Stereographic
            Equal-Area: Oblique Lambert Azimuthal Equal-Area
    E. Straight rhumb lines (principally for oceans)
        Mercator
    F. Straight great-circle routes
        Gnomonic (for less than hemisphere)
    G. Correct scale along meridians
        (1) Centre at pole
            Polar Azimuthal Equidistant
        (2) Centre along Equator
            Plate Carrée (Equidistant Cylindrical)
        (3) Centre away from pole or Equator
            Equidistant Conic

---

properties have been omitted, especially those related to the geometry of the graticule (see section 6.2.1).

Nyerges and Jankowski (1989) adopted Snyder's scheme to develop a knowledge base for map projection selection. The formalised knowledge was implemented in a prototype expert system for map projection selection, known as MaPKBS or *Map Projection Knowledge-Based System* (Jankowski and Nyerges, 1989). *Factual knowledge* in MaPKBS relates to the geographic area to be represented, and is formalised in the form of hierarchically structured triplets *category – object – attributes*. Categories include world, hemisphere and region, as well as continent, ocean, sea and country. The first three categories are *ambiguous*, and the number of objects belonging to each of these categories is indefinite. By changing the aspect, the location of the centre and the specification of the boundaries an infinite number of "worlds", hemispheres, and regions can be defined. The other four categories are *unambiguous*. The number of objects belonging to each is finite. Each object corresponds to one spatial entity and is characterised by a set of geographic attributes. Following Snyder's scheme, two geographic attributes are defined for each object, i.e. location and directional extent. For unambiguous categories, the factual information about objects is expressed in *frames* that are stored permanently in the knowledge base. Each frame contains object's name, category name, location and directional extent. For ambiguous categories, factual information is obtained by querying the user. The information is temporarily stored in the *fact base* and is disposed when the consultation session ends.

The *operational knowledge* of map projection selection describes the relations among object attributes, properties of a map projection, map use and specific map projections. This information is represented in *production rules* of the form IF ..., THEN ..., which are grouped into chains of reasoning paths, each leading to a particular projection. Map projection selection is implemented as a process of concept specialisation. Starting with the definition of the geographic area category, information concerning category object, geographic attributes, map function, geometric properties of the projection, type of display, and map scale is progressively derived through reasoning, using production rules and factual knowledge that is provided by the user or extracted from the frames. Figure 6.1a indicates the successive steps in the process of information gathering needed to solve the task of map projection selection, as defined by Jankowski and Nyerges (1989). Figure 6.1b presents the representational structure of the knowledge base. As can be seen, object information can be easily added to the knowledge base by the creation of new frames. This is accomplished automatically by the system.

MaPKBS, as presented by Jankowski and Nyerges in 1989, cannot be regarded as the ultimate solution to the problem of automated map projection selection. It is a prototype expert system that takes into account only a fraction of the criteria that may be involved in the selection process. At the time of publication, the choice was restricted to conformal and equal-area projections. However, the development of the system proves that the trade-offs involved in map projection selection can be managed by an automated system, if appropriate heuristics are defined, with or without the use of AI techniques. Snyder (1993, p. 276) mentions two other attempts to develop expert systems for map projection

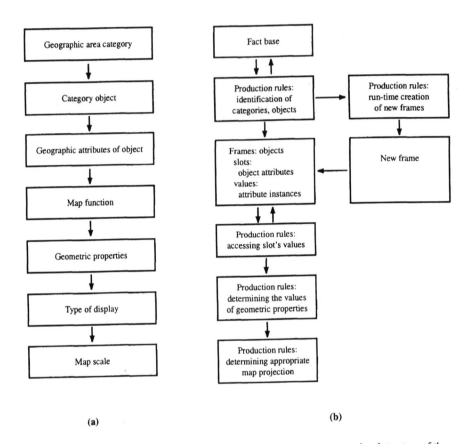

(a)   (b)

Figure 6.1 Process of information gathering in MaPKBS (a), and representational structure of the knowledge base (b) (Jankowski and Nyerges, 1989).

selection, one by Smith and Snyder (1989), and one by Kessler (1991). The first attempt has not been described in sufficient detail to review it. The second has never been officially published. According to Snyder, MaPKBS as well as the other two attempts have been aborted in the research stage because the principals became involved in other projects.

Mekenkamp (1990) presents a very simple procedure for automated map projection selection that is. based on a set of no more than eleven projections, all belonging to the conic group (table 6.3). For Mekenkamp the selection process consists of answering two fundamental questions: (1) what is the shape of the region to be mapped?, and (2) what is the purpose of the map? He distinguishes between *one-point* (round), *two-point* (rectangular), and *three-point* (triangular) regions, leading to the choice of an azimuthal, a cylindrical and a conical projection respectively. Mekenkamp only considers oblique aspects, since these produce the least distortion for a given area. By letting the position of the meta-pole move

Table 6.3 Mekenkamp's set of conic type map projections (Mekenkamp, 1990).

AZIMUTHAL (ONE-POINT AREA)	
1. CON	Conformal
1. EQV	Equivalent
1. EQD	Equidistant
1. GNO	Gnomonic
1. ORT	Orthographic
CYLINDRICAL (TWO-POINTS AREA)	
2. CON	Conformal
2. EQV	Equivalent
2. EQD	Equidistant
CONICAL (THREE-POINTS AREA)	
3. CON	Conformal
3. EQV	Equivalent
3. EQD	Equidistant

without any restriction, the general location of the area (near the pole, near the equator, at mid-latitude), which is one of the main criteria in Snyder's scheme, becomes irrelevant to the selection. This strongly simplifies the selection process. It is up to the user to select the single point that will become the centre of the azimuthal projection, the two points through which the standard circle of the cylindrical projection will run, or the three points that will define the standard circle of the conical projection. Mekenkamp's rule for determining the class of projection is therefore identical to the general rule for the selection of conic type projections that has been formulated at the beginning of this section. It should be pointed out, however, that Mekenkamp also considers a functional argument in deciding on the type of region. Indeed, if attention has to be focused on one point (e.g. the location of an airport), or if the relation between two points is to be emphasised (e.g. the traffic flow between two cities), then the region to be mapped will be designated as a one-point region or a two-point region respectively, and the projection class will be chosen accordingly.

Next to the type of region, the user must select the map property that best suits the purpose of the map. For each class, a choice has to be made between a conformal, an equal-area or an equidistant projection. The azimuthal gnomonic and the azimuthal orthographic projection have been added for special purposes, the first for depicting great circles as straight lines, the second for representing the Earth as seen from a very far distance. Mekenkamp claims that his set of eleven map projections is sufficient for the production of small-scale maps with moderate distortion, at least if the aspect can be chosen freely. The procedure is straightforward, it can be easily implemented, and it always produces a unique solution, which is not so for the decision model proposed by Snyder (table 6.2). Still, the method has some important restrictions. First of all, it is not always easy to characterise a region as extending equally in all directions, extending along a great circle, or extending along a small circle. This was clearly demonstrated in one of the case studies of the previous chapter (see section 5.4.1). Hence, one will not be certain that the projection class that is selected is an appropriate choice in terms of distortion characteristics. Especially for large and/or fragmented areas (e.g. different landmasses) abstraction of shape (round, rectangular, triangular) is difficult. A second problem is the restriction to conic type projections. The simple geometry of the conic group implies that Mekenkamp's approach cannot be used for global mapping purposes without introducing excessive distortion. Finally, since the choice is restricted to oblique aspects of conic projections, the user is not allowed to specify geometric properties that dictate the position of the meta-pole and/or the type of projection, e.g. a straight central meridian, a straight equator, straight parallels, ... While geometric properties are seldom mentioned in connection with map projection selection, they prove to be very important criteria that are often applied in cartographic practice, although mostly the map maker is not conscious of the fact that he or she is actually applying them (see section 6.2.1).

All approaches to map projection selection that have been described so far are strictly occupied with the identification of map projection class (and aspect), and map projection properties. While decisions on map projection class and aspect are clearly made with the intention of reducing distortion, the issue of minimising distortion is seldom raised in map projection selection. Snyder (1987a, p. 34), for example, points out that once a map projection has been selected its parameters must be determined, and that this can be done by estimation or by using more refined methods to reduce distortion. Yet he does not consider optimisation of the projection parameters as a part of the selection procedure itself. Very often, however, map projection selection produces not one, but a set of candidate projections, all of them fulfilling the requirements imposed by the application. In those cases, minimisation of distortion may be useful to decide which projection is best. In the simple schemes that have been presented so far the occurrence of more than one candidate projection is an exception. However, each of these schemes considers only a small subset of all possible map projections. Also, the number of map projection properties is restricted to an absolute minimum. If all map projection properties that are useful in small-scale map design must be dealt with, and if global, continental, as well as regional mapping are considered, it will be impossible to come up with a unique candidate projection, just by looking at the properties the map projection should have. In such case, choosing the projection that has the least distortion for the area to be mapped seems the most obvious way to solve any ambiguity that may occur.

Bugayevskiy (1982) is the only reference known to this author that explicitly discusses the use of quantitative distortion criteria in automated map projection selection, although his approach is purely theoretical (see also Bugayevskiy and Snyder, 1995, pp. 259–61). Once a set of candidate projections satisfying all essential properties has been obtained, he suggests to compute for each projection the weighted mean of different integral measures of distortion that can be derived from local distortion theory:

$$E_{GEN} = \sum_{i=1}^{n} P_i E_i \left/ \sum_{i=1}^{n} P_i \right. \tag{6.1}$$

with E_i an integral measure of distortion, and P_i the corresponding weight factor, and then choose the projection for which the generalised measure is a minimum. Bugayevskiy presents a whole list of potentially useful distortion measures, and indicates that the significance of each measure in the weighting should depend on the purpose of the mapping. Yet he does not offer an objective means to determine the relative weight of each measure. This would present a major difficulty in the practical use of (6.1) in automated map projection selection. It should also be noted that Bugayevskiy's approach is entirely based on local distortion measures and, as such, will probably be less useful for the mapping of large areas.

Maling (1992, p. 239) presents a somewhat simpler approach. Also preoccupied with the need to reduce distortion, he states that the best choice of projection, once class, aspect, and map projection properties have been selected, is the one that provides the least error, when applying the minimum-error principle (see section 5.1). Maling admits that in practice the use of minimum-error projections is the exception rather than the rule. A major reason for this is the fact that until very recently the theory of minimum-error projections was not commonly known to cartographers. Also the optimisation of map projection parameters, or the transformation of map projections to obtain the least possible error, requires sufficient computational resources, which were not available to traditional map makers. Nowadays both obstacles have largely disappeared, and there is no reason why one should not make more use of numerical optimisation methods. Just like Bugayevskiy, Maling considers map projection selection as a process that consists of two steps, i.e. the definition of map projection properties and the minimisation of map projection distortion. Only Maling chooses for one particular distortion criterion, instead of a weighted combination of several criteria. Again it may be argued that the sum of the squares of local scale errors, as used in minimum-error theory, is not an optimum criterion for the evaluation of small-scale maps, and that finite measures of distortion, like the ones proposed in this study, are more appropriate. Using other distortion criteria, however, does not change Maling's basic idea of including map projection optimisation in the selection process. This idea is especially interesting for computerised map projection selection. It has been the starting point for the development of the selection strategy that is presented in this chapter (see section 6.3).

6.2 IMPORTANT FEATURES OF A MAP PROJECTION AND THEIR USE

Any attempt to develop a map projection selection strategy that is to be of practical use has to start with the definition of appropriate selection criteria. Since map projection selection is related to the purpose of the map, one should be able to identify which map projection properties are essential for a particular application. This is not an easy task. Knowledge about the relationship between map application and map projection properties is sparse and sometimes inconsistent. Also, no detailed taxonomy of map applications (map function, map use) is available that could serve as a reference for a systematic study of map projection requirements (Nyerges and Jankowski, 1989). This is the main reason why in the few selection strategies that have been proposed so far, map projection features are the entry to the selection process. It leaves all the responsibility to the user, who is supposed to identify those features that he or she thinks are most relevant to the application. The same approach has been followed in this study, although it is well realised that this will make the selection strategy that will be proposed unsuited for non-experienced cartographers and map users. More research on the relationship between map application and map projection properties, including the integration of already existing knowledge, is highly needed in order to develop methods that can assist the non-experienced user in choosing the right projection. It is beyond the scope of this study to enlarge upon this subject.

Although no comprehensive treatise on the map projection features that are relevant to different types of applications exists, many textbooks on cartography and map projection provide a list of map projection properties, sometimes with examples of their use. Some authors have explicitly addressed the role of projections in map design. Important contributions to the subject are the paper by Hsu (1981), and the collaborative effort of the *Committee on Map Projections* of the ACA (1991), both illustrating a number of ways to make map projections serve different purposes. The following summary of the most useful features of a map projection is largely based on these two references, as well as on our own experience with map projection optimisation. The features that are listed below will form the basis for the selection strategy that will be presented (see section 6.3). A distinction will be made between geometric features, which refer to the geometry of the graticule, and special features, which are related to the distortion characteristics of the projection.

6.2.1 Geometric features

Although most classification schemes emphasise the role of special distortion properties, and pay less attention to the appearance of the projection, some geometric features of the graticule are important in the selection process. To explain the role of geometric features in map projection selection, it is best to distinguish between two types of geometric features, i.e. those that are imposed to serve the purpose of the map, and those that are imposed to constrain the geometry of the graticule when projections are transformed. The latter are only important if the selection process includes a phase of optimisation, in which the shape of the graticule is modified to reduce overall distortion. As has been shown in this study, a well-considered choice of geometric constraints is often a prerequisite to obtain optimised projections with a well-balanced pattern of distortion, especially for the mapping of large areas (see section 5.3). We may refer to the first type of geometric features as *functional features*, and to the second type as *distortion-related features*. It is important to note that most geometric features may serve both purposes, and should not be considered as strictly belonging to one type. No attempt will therefore be made to arrange the features that will be discussed in two distinct groups, yet for each feature its possible use as a functional or a distortion-related feature will be explained.

Outline of the map: the outline of the map may help to focus on a specific feature of the area to be represented. A circular outline is said to give a good impression of the spherical shape of the Earth (Dahlberg, 1991). Although less suited for world maps, it is often used for the mapping of a hemisphere. Portraying the Earth in two hemispheres may be useful to show interesting features of the Earth, for example the division in a "land" and a "water" hemisphere (Olson, 1991), or the position of the "old" and the "new" world. A rectangular outline, as obtained with cylindrical projections, has the advantage of fitting well in the format of a piece of paper (Dahlberg, 1991), although many cartographers plead not to use cylindrical projections for world maps because of the extreme amount of distortion (ACA, 1989). When other than cylindrical projections are used for world maps, a close approximation of the rectangular shape may still be considered important since it reduces the amount of white space on the paper. Maps with strongly curved meridians are often shown without their extremities, to represent the continental areas at the maximum scale. For global maps that represent the pole by a point one often prefers an oval outline, since it gives a good impression of the spherical shape of the Earth. This is especially so for oblique aspects of world maps which are used to simulate a perspective view (see section 6.2.2). Outline is especially important for world maps and hemispherical maps. Maps of other regions are conveniently cut to fit the format of the map sheet. Sometimes also on a regional map the circular outline may be preferred, mostly to draw attention to the centre of the projection, for instance, when using an azimuthal projection to map the radial spread of a phenomenon. However, it must be said that in most examples of this type of mapping, the area is also extended to a hemisphere, or even to the entire world (see also section 6.2.2).

Shape of the parallels and the meridians: in earlier days, map projections that could be easily constructed by graphical means, without the need for computation of coordinates, were largely preferred. As has been said before, the popularity of the transverse azimuthal stereographic projection in the seventeenth century was probably due to its simple construction. Some of the earlier map projections with rather complicated transformation formulas were originally developed by graphical construction (e.g. Ortelius, van der Grinten). Very often these projections have straight or circular meridians and parallels. According to Steers (1970, p. 217) "... it is a good argument that if two graticules are equally suitable for a particular map, it is common sense to draw the easier one". Of course, with the automated plotting of map graticules, the ease of drawing is no longer a valuable argument. For some applications, however, the shape of the parallels and the meridians may still be an important feature. Straight parallels are often preferred for the mapping of phenomena that vary with the geographical latitude (e.g. climatic zones), since they make the association between the spatial pattern of the phenomenon and the latitude more obvious (Hsu, 1981). For the same reason, cylindrical projections are usually preferred for the mapping of phenomena that vary with the longitude, e.g. the boundaries of standard time zones (Robinson, 1991). Cylindrical projections are also advisable for the mapping of north–south traversals of satellites, since they emphasise the cyclical pattern of the ground track (Hsu and Voxland, 1991, see also 6.2.2). Even if all the meridians and the parallels are curved, the map maker generally prefers a straight central meridian (and for world maps a straight equator), in order to obtain a more regular graticule. The choice of straight central axes mostly goes with the demand for graticule symmetry (see below).

Symmetry of the graticule: absence of symmetry in a map projection is often experienced as confusing and unattractive. In school atlases, as well as in the media, almost all maps show a high degree of symmetry, partly to allow non-experienced map users to orient themselves better, partly for aesthetic reasons. We can distinguish between various types of symmetry, the most important being radial, single (along one axis) and double (along both axes) symmetry. Most world map projections in their normal aspect are symmetrical about both the equator and the central meridian. Map projections for regional use are usually symmetrical about the central meridian. For world maps designed for secondary-school students, the central meridian is mostly chosen so that the relative positions of the continents are conventional, with Asia situated in the eastern part of the map, and North and South America in the western part (Bugayevskiy and Snyder, 1995, p. 237). The use of oblique aspects that are not symmetrical about a central meridian (for regional maps), or do not have a doubly symmetrical graticule (for world maps), is still fairly uncommon. Oblique world maps are only used for special purposes (e.g. the representation of major air routes), or to emphasise a certain region by putting it in the central, least distorted part of the map. Symmetry conditions do not only have a functional role in map projection selection. Since they have a strong impact on the regularity of the graticule, they are very useful in the derivation of transformed graticules with balanced patterns of distortion. Furthermore, they substantially reduce the complexity of the optimisation process (see section 5.3.2, *Low-error polyconic projections with pole line*). The advantages of imposing symmetry also hold for the transformation of oblique aspects of projections, although the symmetry of the meta-graticule will only be noticed along the edges of the map (see section 5.3.2, *Low-error oblique projections*).

Spacing of the parallels and the meridians: an equal spacing of the parallels and the meridians along the symmetry axes of the projection contributes to the simplicity of the graticule. Bugayevskiy and Snyder (1995, p. 237) therefore list it as a desirable feature for school maps. The equal partitioning of the parallels by the meridians on the globe is linked to the important concept of geographical longitude. Hence most standard map projections with a straight equator have it equally divided (or nearly so) by the meridians. For the large group of *true* pseudocylindrical projections, which is often used for world maps, each parallel is equally divided by the meridians, just like on the globe. Equal spacing of the parallels along the central meridian is often imposed to avoid extreme compression or stretching in the north–south direction. It produces graticules with intermediate distortion characteristics (less overall scale distortion, compromise between conformal and equal-area projections) (see section 2.5.1). When transforming map projections, maintenance of the equal spacing of the meridians and the parallels along the equator and the central meridian may help to obtain a graticule with a favourable spatial distribution of distortion (see section 5.3.2, *Low-error polyconic projections with pole line*).

Representation of the pole: for the mapping of areas of great longitudinal extent the portrayal of the pole as a point leads to a substantial distortion of shape near the edges of the map. To obtain maps with less extreme distortion of shape along the east and west boundaries, numerous projections with a pole line have been proposed (see section 2.5.2). The lengthening of the pole line, without changing any other characteristic of the graticule, decreases angular distortion at the expense of an increased distortion of areas, especially in the higher latitudes. The increased distortion of area can be diminished by changing the spacing of the parallels along the central meridian, bringing them closer together towards the poles. This will result in a strong variation of scale in the north–south direction, and may lead to a severe stretching of the landmasses in the lower latitudes (see section 2.5.2). Most favourable distortion patterns seem to be obtained for projections with a pole line that is half or nearly half the length of the equator.

Ratio of the axes: although the equator and the meridians are in a ratio of 2:1 on the globe, it has been shown that equatorial projections with a central meridian that is somewhat longer than half the length of the equator have less overall distortion than projections on which both axes are in a correct ratio (see section 2.5.4). However, if one deviates too far from the correct ratio, the stretching of the landmasses in one of the two major directions may be very disturbing to the map reader. As has been said already, small changes in the shape of the landmasses the map reader is familiar with (e.g. Africa) are easily noticed. The ratio of the axes may also be a practical consideration. In many cases the overall shape of the projection must suit a given format. The ratio of the axes and the shape of the parallels and the meridians (see above), are the most effective variables to control the dimensions of the graticule. Both may be modified to obtain a representation of areas at the maximum scale, sometimes at the cost of a less favourable distribution of distortion. It is a good example of the kind of trade-offs that are involved in the process of map design.

6.2.2 Special features

In all textbooks on cartography or map projection theory, conformality, equivalency and equidistance are considered as the most important map projection properties. Next to these three well-known properties, many other useful and less useful features related to map projection distortion have been defined. Some refer to local distortion conditions that are valid in each point on the map, others to local distortion conditions that are only satisfied along one (or two) particular curve(s). Finally, there are also special properties that, even though some of them are locally defined, have a more global meaning to the map projection user, because they refer to the way finite-sized features (great circles, small circles, rhumb lines, satellite orbits, or the graticule as a whole) are represented on the map, or because they indicate a special kind of relationship (distance, direction) between one point (or two points) on the map and all others. We will consider these "global" properties as a separate category.

Some properties are found in most textbooks on map projections, and have well-established names in each major language. Others are more obscure and are not properly defined in each language (Maling, 1968). Maurer (1935), in his study on map projection classification, defines a great number of map projection properties. Only few of these properties are of any practical use. Most of them have no equivalent term in any other language (see also section 1.4.2). Most of the special properties that are frequently found in detailed treatises on map projection in English, German and French are summarised below. If no English term exists, the German (G) or French (F) term is given. The list is not all-inclusive. Less-known properties that have no apparent cartographic utility have been omitted.

Common adjectives for the group of locally defined special features include:

1. Conformal: no distortion of angles, the particular scale is equal in all directions,
2. Equal-area: no distortion of area, the product of the particular scales along the principal directions equals one.
3. Equidistant: the principal scale along the (meta-)meridians equals one.
4. *Abweitungstreu (breitenkreistreu, äquiparallel)* (G), *automécoique* (F): the principal scale along the (meta-)parallels equals one.
5. *Maßähnlich* (G): the principal scale has a constant value along the (meta-)meridians, or along the (meta-)parallels.
6. *Abstandsgleich* (G): (meta-)parallels are straight and equally spaced, but the principal scale along the (meta-)meridians varies.
7. Orthogonal: (meta-)meridians and (meta-)parallels coincide with the principal directions.

The second group of special features of a map projection refers to the same properties as the first. Yet instead of being valid in each point of the graticule, the property is only satisfied along (a) particular curve(s), e.g. along the central meridian, along the equator, along an arbitrary great circle, along one or two arbitrary parallels, or along a user-defined path. To refer to these features one may simply add the suffix "along ..." to the adjective indicating the property that applies, e.g. "equidistant along the central meridian". If a projection has one or two lines of zero distortion, it automatically implies that the projection is conformal, equal-area, equidistant in all directions, and orthogonal in each point along these lines. Some map projections do not have a line of zero distortion, but show one great circle (usually the (meta-)equator) true to scale. By properly re-orienting the graticule of these projections, they may be used for representing the shortest distance between two arbitrarily selected points by an equidistant, straight line (see below). Some map projections have been especially designed for the way they represent a particular curve on the map. A good example is the *Space Oblique Mercator* (SOM) projection, which maps the scanned swath of a satellite nearly conformal throughout all orbits cycles, with the groundtrack continuously at correct scale (Snyder, 1978b). Other examples are the *Generalized Equidistant Cylindric* and the *Generalized Equidistant Conic* projection, which may be used to map one or two paths that do not follow any fixed parallel conformally and correctly scaled (Strebe, 1994). Strebe presents an example of a generalised equidistant cylindric on which the ecliptic is conformal and true to scale, as well as an example of a generalised equidistant conic that preserves correct angles and scale along two non-intersecting great-circle routes.

Of the special properties with a "global" meaning the most important are:

1. Total area true: the total area of the map equals the area of the generating globe.
2. Orthographic: perspective projection from infinite distance (globe-like view).
3. Perspective: general perspective projection onto a horizontal or tilted plane.
4. Stereographic: all great circles and small circles are shown as circles.
5. Orthodromic: all arcs of great circles are shown as straight lines.
6. Loxodromic: all rhumb lines are represented by straight lines.
7. Satellite-tracking: satellite groundtrack is represented by a straight line.

8. Azimuthal: azimuthal angles at the centre of the projection are shown correctly, great-circle distances from the centre are shown as straight lines.
9. Retroazimuthal: the azimuth from every point on the map to the centre of the projection is shown correctly.
10. Two-point azimuthal: azimuthal angles at two given points are not distorted.
11. Two-point equidistant: the distance from every point on the map to two selected points is correct.

While many of the locally defined properties discussed above are shared by a large number of map projections, most global properties are only met by one single projection or by a few projections that have been especially designed for that purpose. Well-known examples are Mercator's projection (loxodromic), Craig's Mecca projection (retroazimuthal) (Craig, 1910), Snyder's projections for the mapping of Landsat groundtracks (satellite-tracking) (Snyder, 1981a), the two-point azimuthal projection (orthodromic, two-point azimuthal) (Maurer, 1914), and the two-point equidistant projection (Maurer, 1919b; Close, 1921).

We will now briefly report on the suggested use of each map projection feature as it is documented in general textbooks on cartography and map projection:

Conformality: in a conformal projection all angles measured from a point on the map are correct. The feature is of great importance for navigational charts, and may also be useful if direction is shown by arrows (e.g. on weather maps), or if the shapes of isolines are of major importance (Bugayevskiy and Snyder, 1995, p. 244). Distortion of area on conformal map projections is extreme. Hence it is not advisable to use conformal map projections for the small-scale mapping of large areas, unless the phenomenon that is to be depicted compels the map maker to do so.

Equal-area property: an equal-area projection has no distortion of area, meaning that all areas on the Earth's surface are represented in correct proportion. It is a property that is considered important for all sorts of small-scale thematic maps that use area shading to show the extension of different thematic categories (area class maps) or the magnitude of a variable within the boundaries of a set of well-defined areas (choropleth maps). Equal-area projections are absolutely required for dot maps to ensure a correct impression of the density of the phenomenon that is being mapped. For the mapping of very large areas (especially world maps) the use of equal-area projections should be avoided, because of the extreme distortion of shape. In such case, one can better use one of the many map projections with low area distortion that are presently available. If the equal-area property is considered indispensable, one might consider the use of an interrupted projection.

Equidistance: equidistance indicates the absence of scale distortion and can only be achieved in one direction, or along specific lines. The term *equidistant projection* refers to those projections that have no scale distortion in the direction of the meridians (or meta-meridians). Especially the oblique azimuthal equidistant projection is of particular use since distances from the centre of the projection to every other point on the map are shown correctly. We will refer to this special property as *radial equidistance*. It makes the projection very suited for maps that focus on one particular point, for example, touristic maps, or maps from airline companies showing correct distances from one city to various destinations (Monmonier, 1991), or seismic maps showing distances from a single seismic station (Bugayevskiy and Snyder, 1995, p. 244). Muehrcke (1991) gives an example of the clever use of an oblique azimuthal equidistant projection to estimate the average time it will take for a tsunami, caused by an earthquake somewhere in the Pacific basin, to reach various islands and continental coastlines.

The property of equidistance is also very useful when the relationship between two locations needs to be emphasised. In that case one will usually prefer to represent the great-circle arc between the two chosen locations by an equidistant straight line. For that purpose any projection showing at least one great circle correctly can be used. It suffices to choose an oblique aspect so that the equidistant great circle of the projection passes through both focal points. Gilmartin (1991) gives examples of oblique cylindrical projections on which the points of interest are aligned along the meta-equator or along an equidistant meta-meridian. Many of the doubly symmetric map projections that are used for world maps today are equally suited for the purpose. We will name this property *equidistance between two points* to distinguish it from *equidistance sensu stricto* (along the meridians), and *radial equidistance*. As indicated in the above list, equidistance along the parallels may also be considered a distinct property, and is even given special names in German and French textbooks. Yet the property has no practical utility, and is therefore not suited to be included in a selection scheme. The same can be said about the maintenance of a constant scale factor along the meridians or along the parallels. In practice this property, which can be regarded as a generalisation of equidistance, is only of use when it refers to the central meridian and the equator of the projection, since it guarantees an equal spacing of the parallels and the meridians along these axes. Applied in that sense, the property is already included in the list of geometric features (see section 6.2.1).

Orthogonality: although a deviation of several degrees from a right-angled intersection of meridians and parallels can be detected easily, small deviations of orthogonality are not considered as particularly disturbing by map users (Mackay, 1969). Moreover, standard map projections with orthogonal graticules (e.g. the group of conic projections) have the disadvantage of being not sufficiently pliable to be adapted to regions of arbitrary shape without renouncing the property of orthogonality. Apart from applications that demand conformality (which automatically implies orthogonality of the graticule), strict orthogonality has no obvious use in cartography. It should therefore not be regarded as an important property for map projection selection. If selection includes an optimisation phase, aimed at reducing the overall distortion of shape on the chosen projection, the presence of sharp graticule intersections in the region to be mapped will be minimised anyhow (for examples, see chapter 5).

Total area true: a great number of projections for the mapping of the entire world have been developed with a total area equal to the area of the generating globe. Although it may be worthwhile to mention this property when describing these projections, the feature has no apparent cartographic utility. As such there is no use in including it in a selection scheme. Most projections that are total area true were obtained by a simple re-scaling of a map projection with equally spaced parallels as a first step in the creation of an equal-area graticule (see section 2.5.2).

Orthographic: the use of map projections that portray the Earth as it is seen from space is becoming increasingly popular. The best-known example of a map projection that provides a space-like view, and definitely the most frequently used, is the azimuthal orthographic projection. It projects the Earth onto a plane from infinite distance, and produces an image of one hemisphere that looks very much like it appears on the globe (Dent, 1987; Brinker, 1990). The globe-like view is especially convincing for the oblique aspect of the projection, with the centre somewhere between the equator and the pole, although the equatorial aspect with its straight parallels is also very often encountered. Compression near the edge of the map is very high, which makes that the projection is only used for pictorial purposes.

Perspective: outer-perspective azimuthal projections with the point of projection close to the Earth's surface may be used to *zoom in* on a particular area or create an effect of *looking beyond* an area (Hsu, 1981). Projecting on a tilted plane instead of a horizontal plane increases the photo-reality of the representation, since the mapping then becomes identical to the mathematical model for ideal aerial and space photographs (Snyder, 1981b; Deakin, 1990). Occasionally oblique aspects of world map projections with a circular or oval perimeter are used to simulate a global view, although these projections are not truly perspective. This may be done, for example, to show all areas in true proportion, which is not possible with any perspective projection (Castner, 1991).

Stereographic: the azimuthal stereographic projection has the interesting property that all circles on the globe are shown as circles on the map. This makes the projection especially suited to illustrate the influential range of phenomena that spread outward in circular fashion from a source area. While any azimuthal projection can be used if only one source area is involved, by simply putting the source location in the centre of the map (see below), the stereographic projection is the only map projection that can be used if more than one source area is involved (Muehrcke, 1991). Possible applications include the visualisation of the potential risk zones of various nuclear plants, the range of radio waves emitted from different ground stations or the radius of action of aircraft (Robinson *et al.*, 1984, p. 100).

Orthodromic: the azimuthal gnomonic is the only projection that represents all great circles (orthodromes) by straight lines. This property makes the projection well suited for great-circle navigation along many routes. However, scale distortion away from the centre of the projection increases very rapidly, and the projection can only be used for areas smaller than a hemisphere. This explains why for great-circle navigation along a single route the oblique Mercator projection is usually preferred (Bugayevskiy and Snyder, 1995, p. 246). Next to navigation, the gnomonic projection may be used for any other map on which various great-circle routes are to be represented by straight lines, e.g. commercial maps of airline companies. Of course, any affine transformation of the gnomonic will maintain the orthodromic property. Maurer presented such an affine transformation that has the additional property of showing all azimuthal angles from two focal points (instead of one) correctly (Maurer, 1914) (see below).

Loxodromic: for rhumb-line navigation the normal aspect of the Mercator projection is used almost exclusively, since it is the only projection that shows rhumbs (loxodromes) as straight lines. The loxodromic property has no apparent cartographic utility outside the domain of navigation, although the normal Mercator projection is of course often used as an educational aid to explain the difference between rhumb-line and great-circle routes to students.

Satellite-tracking: in 1981 Snyder described a new series of conic-type map projections on which inclined circular orbits are shown as straight lines, and called them *satellite-tracking* projections (Snyder, 1981a). He presented several examples of projections for the mapping of Landsat orbits. All projections are conformal along two parallels. For the cylindrical projection (for large areas) both parallels are made true to scale, for the conical projections (for continents and countries) only one parallel is represented in correct length. Although Snyder's map projections facilitate the plotting and locating of satellite groundtracks, they have not gained much popularity. World orbit maps that are found in textbooks on remote sensing are still mostly plotted on standard cylindrical projections, with the groundtracks represented by curved lines. The reason for this may be the extreme distortion of the polar areas on Snyder's cylindrical satellite-tracking projection, and the fact that areas beyond the limits of latitude reached by the groundtrack cannot be shown.

Azimuthal: azimuthal projections have a radial distortion pattern and are therefore optimally suited for the representation of phenomena with a circular extent. Any circle on the globe that is centred on the point of zero distortion of the projection will be preserved on the map. Muehrcke (1991) used an oblique azimuthal projection centred on the southern tip of Africa to show the *Ring of Fire*, i.e. the circular band of tectonic activity surrounding the Pacific Ocean and spreading outward from the position of the original ancient land mass. Hsu and Voxland (1991) suggested the use of an azimuthal projection to emphasise the circular nature of the flight of Voyager, and to show the near-equatorial flight path around the world without any interruption. Azimuthal projections are useful for showing the direction in which to point antennas to receive radio signals from any station. When using an equidistant azimuthal projection not only bearings, but also the distance the radio waves have to travel to reach their destination, can be read immediately from the map.

Retroazimuthal: on retroazimuthal projections the azimuth from every point on the map to the centre of the projection is shown correctly. Just like azimuthal projections they can, for example, be used to point antennas to receive radio signals, yet this time the source of emission instead of the receiver is put in the centre of the map. The first retroazimuthal projection was developed by Littrow in 1833, but had not been designed for that purpose. The projection is conformal and shows correct azimuths to any point on the central meridian, yet this feature of the projection was only discovered in 1890 (Snyder, 1993, p. 227). The best-known retroazimuthal is probably the *Mecca* projection, designed by Craig (1910) to allow worshippers in Islamic countries to determine the direction to face for prayers. Other retroazimuthal projections have been proposed by Hammer (1910), Maurer (1919a), Arden-Close (1938), Jackson (1968), and Hagiwara (1984). Except for Hammer's projection, which combines correct azimuth from all points to the centre with radial equidistance, all these projections can be considered as purely academic efforts. As Snyder points out, the azimuthal stereographic projection, which is conformal, can equally well be used to determine the azimuth from every point to the centre of the projection. Yet instead of measuring the angle from the vertical, it must be measured with respect to the direction of the meridian on which the point is located (Snyder, 1993, p. 231).

Two-point azimuthal: the two-point azimuthal projection has correct azimuths from two points (instead of one) and represents all great-circle arcs by straight lines. It is obtained by affine transformation of an oblique azimuthal gnomonic projection of which the centre is located halfway between the two points from which correct azimuths are desired. The projection was first developed by Maurer (1914) and reinvented by Close (1922) (see also Young, 1922). It can be used for plotting radio waves arriving at two stations, and for finding the source of the waves (Steers 1970, p. 192).

Two-point equidistant: next to his two-point azimuthal and two-point retroazimuthal, Maurer proposed a two-point equidistant projection in which all distances measured from two chosen points are correct (Maurer, 1919b). Just like the two-point azimuthal it was independently reinvented by Close (1921), who also proposed a third variant on which distances are correct from one point and azimuths are correct from the other (Close 1922, see also Young 1922). The two-point equidistant projection was adopted by the *National Geographic Society* (NGS) of the United States for maps of Asia, not to measure distances from two points, but to reduce overall distortion (Snyder, 1993, p. 234). For maps of the other continents the NGS developed its own projection, the *Chamberlin trimetric*, which is an approximate three-point equidistant projection (Bretterbauer, 1989; Christensen, 1992). Since both the two-point equidistant and the Chamberlin trimetric are used for their low distortion of finite distances they can be considered as precursors of the projections with low finite distortion that have been proposed in Peters (1975), Tobler (1977), as well as in this study. The major difference is that in the latter studies not two or three points, but a large set of points is used to reduce overall distortion, and that projections are derived numerically.

6.3 SEMI-AUTOMATED MAP PROJECTION SELECTION

6.3.1 General principles

Based on our knowledge about existing selection schemes (see section 6.1), and the criteria that are important in map projection selection (see section 6.2), a strategy for semi-automated map projection selection will be proposed. In developing this strategy the following seven principles have been taken into account:

1. The selection method should be based on the purpose of the map.
2. The system should provide a unique solution for any application.
3. The selected projection must have low overall distortion for the area to be mapped.
4. The method must be applicable to global, continental and regional mapping.
5. The method should work for each set of map projection features.
6. The user should be able to evaluate and possibly reject the selected projection.
7. The system should make optimum use of the possibilities offered by working in a digital environment.

When setting up a selection system one should first of all decide which features are to be included, and then define a set of map projections that covers all possible combinations of these features, or at least those that are relevant. At this point it must be noted that not all features that are included in the selection can be combined. Pseudocylindrical projections, for example, can never be conformal. On the other hand, some combinations of features may be theoretically possible but are not desirable. Although a conformal world map with a circular outline can be produced, overall distortion on such a map would be so large that it may be considered useless. The choice of map projections to be included in the selection scheme is self-evident for those features that are unique to one single projection. If the loxodromic property is considered important, then it is obvious that Mercator's projection has to be included. Most features, however, are not unique to one single projection, and make it more difficult to decide which projections are to be included, and which are not. If one wants to keep the total number of map projections to a strict minimum, in order to reduce the complexity of the decision process, it is advisable to work with projections that can satisfy multiple combinations of features. Wagner's general projections, which have been discussed in detail in chapter 4 (see section 4.1.2), are very interesting in this respect, since the geometric properties of these projections can be adjusted by changing the values of the parameters in the transformation formulas. Using these map projections, various combinations of properties can be handled by one and the same transformation.

Map projection selection should start with the definition of the geometric features and special features that are thought to be essential for the application one has in mind (see section 6.2). In an ideal situation the system should be able to identify the required features automatically from the type of application, or assist the user in choosing the proper features. This was not pursued. As has been said before, the relationship between map application and map projection features is far from fully understood. Knowledge-based approaches to map projection selection are seriously hampered by problems related to the formalisation and integration of existing knowledge on map projection use, and have been limited in scope (see section 6.1). Very often requirements are contradictory, and the selection of interesting map projection features involves various trade-offs that cannot be translated into formal rules. Hence it will be assumed that the user himself is able to decide which properties are best selected. This implies that the method that is proposed will only be useful for those who have at least some knowledge of map projection principles. It is very doubtful if it will ever be possible to develop a selection tool that does not require some knowledge on the part of the map maker. Unless a very simple method with limited options is concerned that offers sub-optimal solutions, or that is only applicable within a certain range of scales.

Specifying the features that are required will reduce the number of candidate projections, but will not necessarily produce a unique solution. Depending on the characteristics of the projections that are considered in the selection process, and the features that are requested, one may be left with several projections fulfilling all requirements. Limiting the number of map projection classes in relation to the extent of the area to be mapped, or choosing for one particular aspect, will allow a further reduction of the number of candidate projections (see section 6.3.3), but cannot guarantee that an unambiguous choice is made. One way to decide which of the remaining projections is best would be to calculate for each projection mean and maximum distortion values for the region to be mapped, by choosing appropriate values for the parameters of the projection, or by optimising the value of each parameter automatically. Although there is no particular difficulty in applying this method, it is somewhat clumsy and will consume too much time if several candidate projections are involved. It is therefore advisable to use a more rational method of choice that produces a suitable projection without the need for excessive computation. For the selection method that is presented in this study, it was decided to rank all candidate projections using a simple set of rules that favour those projections that have the most balanced pattern of distortion, and the highest flexibility for graticule optimisation. The same set of rules can be applied if the selection procedure is based on a different set of features and map projections.

One of the advantages of working in a digital environment is that the choice of map projection parameters does not have to be based on the use of simplified rules of the type described by Maling (1960, 1992, pp. 241–2) for the choice of the standard parallels of conical projections. Although the user should have the opportunity to define his own projections parameters, the use of numerical techniques to reduce overall distortion will relieve him from the burden of having to select appropriate values for the parameters himself, and will produce an optimal solution. The combination of map projection selection (based on functional requirements), and map projection optimisation (to reduce distortion), is a key feature of the selection procedure that is proposed. Next to simple optimisation of the standard parameters of the projection, also polynomial transformation, which has been explained in detail in the two previous chapters, will be considered. The main reason for including polynomial optimisation in the procedure is that standard projections not always yield sufficiently low distortion, especially not for very large areas, or for areas that do not fit well in the distortion pattern of the projection (for examples, see sections 5.3.2, 5.3.3, 5.4.2, 5.4.3). As has been shown in chapter 4, most of the important features of a map projection that have been listed in section 6.2 can be maintained when polynomial transformation is applied.

Another important element in the method for map projection selection that is proposed is the interaction with the user. In the previous chapters of this study it has been made quite clear that map projection optimisation not necessarily produces an acceptable result. It is therefore up to the user to decide if the proposed solution is satisfactory or not, and which actions should be taken for improvement. A quick visualisation of the graticule of the selected projection, if possible with isocols showing the distribution of distortion, as well as a brief report, listing the optimised value for each projection parameter, the mean overall distortion, and the range of distortion values (or maximum distortion) within the mapped area, is therefore essential. For global applications the user will be mostly interested in the overall appearance of the continents, and seek to obtain a favourable spatial distribution of distortion values. For the mapping of smaller areas the purely quantitative aspects of distortion become more important, and the user may for instance wish to specify certain limiting values of distortion that should not be exceeded.

If parameter optimisation does not produce an acceptable graticule, the user may decide to apply polynomial optimisation, using the same constraints (map projection features) as for the selection of the projection. If this does not lead to a major improvement the user may have to reconsider his feature specifications, release one or more features that are less essential for the application, and restart the selection process. For the mapping of large areas, where the overall distribution of distortion is crucial (see above), it will often be more interesting to restrict the geometry of the graticule in order to obtain a more balanced pattern of distortion, instead of releasing some of the constraints (see section 5.5). If an indirect aspect has been selected, obtaining a balanced pattern of distortion can be achieved by restricting the flexibility of the meta-graticule. The introduction of additional constraints, as just indicated, not necessarily requires the selection process to be repeated all over again. Indeed, the (meta-)graticule of the projection that has been selected may already have more geometric qualities than those that were originally specified by the user, some of which may be useful for controlling the transformation. This will, for example, be the case when the user wishes to optimise an indirect aspect of a doubly symmetric projection without losing the two-fold symmetry of the meta-graticule, a situation that will regularly occur when developing indirect aspects of projections for world maps (for some examples, see section 5.3.2, *Low-error oblique projections*).

Again it should be clear that the trade-offs that are involved in producing a low-error compromise require some basic understanding of map projection principles on the part of the map maker, who needs to provide the system with essential input. Map projection selection is thus seen as a dynamic process, which can only be partly automated and which still requires a great deal of human intervention (Canters, 1995). Procedures for semi-automatic map projection selection like the one that is proposed here will increase the level of objectivity in the choice of an optimal map projection, and will assist the map maker in reaching an appropriate solution, yet they still depend on his expertise. Figure 6.2 schematises possible flows of information in the process of selecting a proper map projection, as seen by this author.

6.3.2   Included features and map projections

The general principles for map projection selection that have been presented above may be the starting point for the design of various feature-based selection procedures, depending on which map projection properties the designer wishes to include. In the following we will propose a procedure for selection that is based on most of the features that have been discussed in section 6.2. Some features have been left out because they are considered less relevant for selection purposes (equidistance along the parallels, orthogonality of the meridians and the parallels, total area true representation of the globe), others because they refer to properties that are only occasionally requested (satellite-tracking, retroazimuthality, two-point azimuthality, two-point equidistance). The procedure, as it is proposed in this study, should therefore be considered only as one particular solution to small-scale map projection selection, using a minimum set of features and map projections, which is thought to be sufficient for the majority of map applications. If more features (and map projections having these features) are required the procedure should be extended. This can be done easily without altering the general approach to map projection selection that is presented.

Table 6.4 lists all features that have been included, as well as the minimum

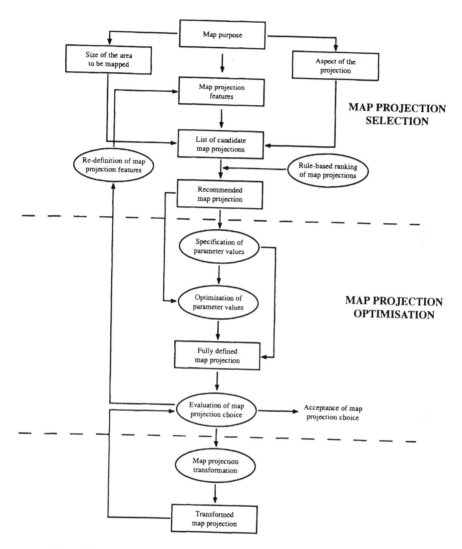

Figure 6.2 Possible flows of information in the selection of a suitable map projection.

set of map projections that have been selected to cover all useful combinations of these features. The set includes the conformal, equal-area and equidistant version of the azimuthal, cylindrical and conical projection, which are sufficient for most continental and regional mapping tasks. The gnomonic, orthographic, and general perspective azimuthal projection, have been added for their special properties. For the cylindrical equidistant and the cylindrical equal-area projection a distinction is made between the version with one and two standard parallels. Although the use of two standard parallels will always yield the lowest overall distortion, the use of one standard parallel will provide equidistance along the equator (or meta-equator). For global mapping purposes, four of Wagner's general doubly symmetric projections have been included. By putting appropriate restrictions on the value of the parameters in the transformation formulas of Wagner's first and sixth projections (see section 4.1.2), most useful combinations of features for the pseudocylindrical type of projection (straight parallels) can be reproduced. In a similar way, the Hammer–Wagner and Aitoff–Wagner projection cover the most interesting combinations of features for the general polyconic type of projection (curved parallels). The Mollweide projection has been added because it is the only pseudocylindrical equal-area projection that has an oval outline and a correct ratio of the axes. Tables 6.5 and 6.6 list parameter values and dependencies for all variants of the four Wagner projections that are included in table 6.4. Each variant corresponds to a particular combination of map projection features.

Table 6.4 can only be used to identify the features of each map projection in its direct aspect. A change of aspect will alter the appearance of the graticule, and will therefore have a strong impact on the properties of the projection. All features of the graticule in its normal aspect will become features of the meta-graticule when the aspect is changed. Geometric properties that refer strictly to the meridians and the parallels may therefore be lost (e.g. parallels and meridians can no longer be shown as straight lines). On the other hand, properties that implicitly refer to the meta-graticule will become more generally applicable by allowing the aspect to change (e.g. equidistance along any selected great circle instead of equidistance along the equator). It must be noted that some geometric properties may relate both to the graticule and the meta-graticule, depending on whether they are applied for purely functional reasons, or to control the optimisation process (see section 6.2.1, 6.3.1). A good example is the equal spacing of the longer axis (meta-equator) on a doubly symmetric projection. This is a quality that may be imposed to obtain a simple normal graticule with an equally spaced equator (e.g. for world maps in a school atlas), but also to obtain an optimised oblique graticule with a well-balanced pattern of distortion (see section 5.3.2, *Low-error oblique projections*).

Since aspect should be taken into account in the selection procedure, decision matrices similar to the one shown in table 6.4 have been compiled for each aspect of the projections that are included in the selection. Tables A.1 and A.2 list all features for the transverse and oblique aspect of the conic-type projections, while tables A.3 through A.7 describe the relationship between features and map projections for the remaining six aspects of the non-conic projection types (see appendix A). As can be seen from the tables, all features that refer to the meta-graticule (including most special features) remain the same for each aspect of a map projection. Only the properties of the regular graticule (meridians and parallels) will change from one aspect to another. Note also that the skew and the plagal aspect are both represented by the same matrix. Although in a purely mathematical sense the skew aspect is a specialisation of the plagal one (see section 2.6), both aspects are identical in terms of the geometric qualities that are included in the selection procedure, at least in the form in which it is proposed here.

As has been said already, the selection method that is presented is designed to work for any set of features and map projections, not only for those that are

Table 6.4 Features for the normal aspect of all map projections included in the selection procedure.

MAP PROJECTIONS	Radial symmetry	Double symmetry	Single symmetry	Equal spacing along the central meridian	Equal spacing along the equator	Straight parallels	Straight meridians	Circular outline	Rectangular outline	Oval outline	Meta-pole is represented by a point	2:1 ratio of the axes	Conformal	Equal-area	Radial equidistant	Equidistant along the meta-equator	Azimuthal	Loxodromic	Orthodromic	Stereographic	Orthographic	Perspective
Azimuthal																						
Azimuthal equidistant	×	×	×	×			×	×			×				×		×					
Azimuthal equal-area	×	×	×				×	×			×			×			×					
Azimuthal stereographic	×	×	×				×	×			×		×				×			×		
Azimuthal gnomonic	×	×	×				×	×			×						×		×			
Azimuthal orthographic	×	×	×				×	×			×						×				×	
Azimuthal outer-perspective	×	×	×				×	×			×						×					×
Cylindrical																						
Cylindrical equidistant		×	×	×	×	×	×		×							×						
one standard parallel		×	×	×	×	×	×		×							×						
Cylindrical equal-area		×	×		×	×	×		×			×		×								
one standard parallel		×	×		×	×	×		×					×								
Cylindrical conformal		×	×		×	×	×		×				×					×				
one standard parallel		×	×		×	×	×		×				×					×				
Conical																						
Conical equidistant			×	×			×									×						
Conical equal-area			×				×							×								
Conical conformal			×				×				×		×									

Table 6.4 continued.

Column groups — **GEOMETRIC FEATURES**: *Graticule* (Symmetry: Radial symmetry, Double symmetry, Single symmetry; Spacing/shape: Equal spacing along the central meridian, Equal spacing along the equator, Straight parallels, Straight meridians); *Meta-graticule* (Outline: Circular outline, Rectangular outline, Oval outline; Meta-pole is represented by a point; 2:1 ratio of the axes). **SPECIAL FEATURES**: *Locally defined* (Conformal, Equal-area, Equidistant: Radial equidistant, Equidistant along the meta-equator); *Globally defined* (Azimuthal, Loxodromic, Orthodromic, Stereographic, Orthographic, Perspective).

MAP PROJECTIONS	Radial symmetry	Double symmetry	Single symmetry	Equal spacing along the central meridian	Equal spacing along the equator	Straight parallels	Straight meridians	Circular outline	Rectangular outline	Oval outline	Meta-pole is represented by a point	2:1 ratio of the axes	Conformal	Equal-area	Radial equidistant	Equidistant along the meta-equator	Azimuthal	Loxodromic	Orthodromic	Stereographic	Orthographic	Perspective
Pseudocylindrical																						
Wagner VI																						
Equidistant meta-equator		x	x	x	x	x										x						
2:1 ratio of the axes		x	x	x	x	x						x										
Equidistant meta-equator + 2:1 ratio of the axes		x	x	x	x	x						x				x						
Meta-pole is mapped as a point		x	x	x	x	x				x	x					x						
equidistant meta-equator		x	x	x	x	x				x	x					x						
2:1 ratio of the axes		x	x	x	x	x				x	x	x										
equidistant meta-equator + 2:1 ratio of the axes		x	x	x	x	x				x	x	x				x						
Wagner I																						
Equidistant meta-equator		x	x		x	x								x		x						
2:1 ratio of the axes		x	x		x	x						x		x								
Meta-pole is mapped as a point		x	x		x	x					x			x								
equidistant meta-equator + 2:1 ratio of the axes		x	x	x	x	x					x	x		x		x						
Mollweide		x			x	x				x	x	x		x		x						

Table 6.4 continued.

Column groups: **GEOMETRIC FEATURES** — *Graticule* (Symmetry: Radial symmetry, Double symmetry, Single symmetry; Spacing/shape: Equal spacing along the central meridian, Equal spacing along the equator, Straight parallels, Straight meridians); *Meta-graticule* (Outline: Circular outline, Rectangular outline, Oval outline; Meta-pole is represented by a point; 2:1 ratio of the axes). **SPECIAL FEATURES** — *Locally defined* (Conformal, Equal-area, Radial equidistant, Equidistant along the meta-equator); *Globally defined* (Azimuthal, Loxodromic, Orthodromic, Stereographic, Orthographic, Perspective).

MAP PROJECTIONS	Radial symmetry	Double symmetry	Single symmetry	Equal spacing along the central meridian	Equal spacing along the equator	Straight parallels	Straight meridians	Circular outline	Rectangular outline	Oval outline	Meta-pole is represented by a point	2:1 ratio of the axes	Conformal	Equal-area	Radial equidistant	Equidistant along the meta-equator	Azimuthal	Loxodromic	Orthodromic	Stereographic	Orthographic	Perspective
Polyconic																						
Aitoff–Wagner																						
Equidistant meta-equator		x	x	x	x											x						
2:1 ratio of the axes		x	x	x	x							x										
Equidistant meta-equator + 2:1 ratio of the axes		x	x	x	x							x				x						
Meta-pole is mapped as a point																						
equidistant meta-equator		x	x	x	x						x					x						
2:1 ratio of the axes		x	x	x	x						x	x										
equidistant meta-equator + 2:1 ratio of the axes		x	x	x	x						x	x				x						
oval outline										x	x											
equidistant meta-equator		x	x	x	x					x	x					x						
2:1 ratio of the axes		x	x	x	x					x	x	x										
equidistant meta-equator + 2:1 ratio of the axes		x	x	x	x					x	x	x				x						
Hammer–Wagner		x	x											x								
2:1 ratio of the axes		x	x									x		x								
Meta-pole is mapped as a point		x	x								x			x								
2:1 ratio of the axes		x	x								x	x		x								
oval outline		x	x							x	x			x								
2:1 ratio of the axes		x	x							x	x	x		x								

Table 6.5 Parameter values and dependencies between parameters for all variants of Wagner's pseudocylindrical projections included in the selection procedure.

Map projections	Parameter values and dependencies
WAGNER VI	three parameters: m, n, k_1
Equidistant meta-equator	$k_1 = \sqrt{m/n}$
2:1 ratio of the axes	$k_1 = m/n$
Equidistant meta-equator + 2:1 ratio of the axes	$n = m, k_1 = 1$
Meta-pole is mapped as a point	$m = 1$
equidistant meta-equator	$k_1 = 1/\sqrt{n}$
2:1 ratio of the axes	$k_1 = 1/n$
equidistant meta-equator + 2:1 ratio of the axes (Apianus)	$n = 1, k_1 = 1$
WAGNER I	two parameters: m, n
Equidistant meta-equator	$n = m$
2:1 ratio of the axes	$n = \dfrac{2 \arcsin m}{\pi}$
Meta-pole is mapped as a point	$m = 1$
equidistant meta-equator + 2:1 ratio of the axes (sinusoidal)	$n = 1$

included in the tables discussed above. However, some rules have to be taken into account when choosing the projections to be included in the scheme. First of all, members of the conic group should be included, since map projection selection for small areas is confined to this type of projection (see below). Secondly, it is assumed that the set of map projections is kept to a strict minimum, meaning that for every combination of features for the same class of projections one should try to include only one projection. This will guarantee that the selection method offers a unique solution. Since the selection process is only based on map projection features and map projection class (map projection class is used as an internal selection criterion, see below), it cannot distinguish between two projections of the same class that have exactly the same features. The rule also implies that if two

Table 6.6 Parameter values and dependencies between parameters for all variants of Wagner's polyconic projections included in the selection procedure.

Map projections	Parameter values and dependencies
AITOFF–WAGNER	four parameters: m, n, k_1, k_2
Equidistant meta-equator	$k_1 = \sqrt{m/n}$
2:1 ratio of the axes	$k_1 = \dfrac{m}{k_2 n}$
Equidistant meta-equator + 2:1 ratio of the axes	$k_1 = k_2 = \sqrt{m/n}$
Meta-pole is mapped as a point	$m = 1$
equidistant meta-equator	$k_1 = 1/\sqrt{n}$
2:1 ratio of the axes	$k_1 = \dfrac{1}{k_2 n}$
equidistant meta-equator + 2:1 ratio of the axes	$k_1 = k_2 = 1/\sqrt{n}$
oval outline	$n = 0.5$
equidistant meta-equator	$k_1 = \sqrt{2}$
2:1 ratio of the axes	$k_1 = 2/k_2$
equidistant meta-equator + 2:1 ratio of the axes (Aitoff)	$k_1 = k_2 = \sqrt{2}$
HAMMER–WAGNER	three parameters: m, n, k
2:1 ratio of the axes	$k^2 = 2\sin\!\left(\dfrac{\arcsin m}{2}\right)\!\Big/\sin\!\left(\dfrac{n\pi}{2}\right)$
Meta-pole is mapped as a point	$m = 1$
2:1 ratio of the axes	$k^2 = \sqrt{2}\Big/\sin\!\left(\dfrac{n\pi}{2}\right)$
oval outline	$n = 0.5$
2:1 ratio of the axes (Hammer)	$k = \sqrt{2}$

projections of the same class satisfy a particular set of properties, but one of them has additional features, then this is the one that should be included in the scheme. As a third rule, the scheme should include at least one projection with equally spaced (meta-)parallels for each projection class, since the equal spacing of the (meta-)parallels is an important criterion in the selection process (see below). Finally, it is advisable to choose projections that offer good opportunities for parameter optimisation, i.e. projections with easily adjustable graticules.

6.3.3   The selection procedure

The method for selection that is proposed consists of two consecutive processes: the selection of a map projection with the properties that are requested by the user, and the optimisation of the projection's graticule in order to minimise distortion within the region to be mapped. Selection starts with the derivation of a list of candidate projections having all requested properties. From this list one projection is selected. This projection may be optimised in several ways, as described in section 6.3.1, depending on the map maker's concern about map projection distortion. The procedure that is applied to derive a list of candidate projections, and to decide which projection will be recommended, corresponds with the flow chart that has been presented in figure 6.2, and is described in detail below. It is partly based on existing knowledge about map projection selection (see section 6.1), uses what is known about the impact of map projection features on the distortion characteristics of a projection (see sections 2.4, 2.5, 2.6), and takes account of the general design principles listed above (see section 6.3.1).

Derivation of a list of candidate projections

The derivation of a list of candidate projections is accomplished in three steps and is based on the following criteria:

1.   The map projection features that are required.
2.   The extent of the area to be mapped.
3.   The aspect of the projection.

The first step in the selection process involves the specification of the features the map projection should have. The decision matrices that have been discussed earlier (see section 6.3.2) list the features of each map projection in all of its aspects, which allows us to identify all projections, or to put it more correctly, all aspects of all projections, that satisfy the set of user-specified requirements. To derive the set of candidate projections it is very important that all aspects of a projection are examined. Indeed, although it is usually true to say that all the properties a map projection has are found in its normal aspect, there are some exceptions. The transverse aspect of the orthographic projection, for example, does have straight parallels, while the normal aspect does not. If the selection of map projections with straight parallels would be based on the properties of the normal aspect only, as listed in table 6.4, the transverse orthographic would have been mistakenly omitted from the set of candidates.

In the second step of the selection process the extent of the area to be mapped is specified. Regions smaller than a hemisphere (*regional*) can be represented with hardly noticeable distortion using the standard projections from the conic group. Hence, other types of map projections are not considered for the mapping of these areas unless some property should be required that cannot be met by the conic group. For maps of one hemisphere (*hemispherical*), projections with a circular outline are preferable. Selection is therefore restricted to azimuthal projections, again unless properties are required that cannot be met by the azimuthal class. For regions extending the size of a hemisphere (*global*) no *a priori* limitations are set on the type of projection. Map projection selection for these areas is only constrained by map projection properties that are considered indispensable for the application.

Once the set of candidate projections has been obtained, the map maker may still wish to restrict the selection to a particular aspect. This will reduce the set of candidates to those projections that have the right properties in the right aspect. It should be clear, however, that the specification of features nearly always limits the possibilities for the choice of aspect, e.g. an equal spacing between the parallels can only be achieved if one of the axes of the map projection coincides with a meridian. Hence the demand for a particular aspect may require that some of the features that have already been defined have to be released. It is up to the map maker to decide if this is justified. In the strategy that is proposed here the request for particular map projection properties is taken into account before the aspect is chosen, since it is our opinion that the properties of a projection are more closely linked to the purpose of the mapping than the choice of aspect. This, of course, should not prevent the map maker from giving more weight to the choice of aspect by re-definition of the list of requested properties. It should also be noted that the map maker is not forced to choose for a particular aspect. If no choice of aspect is made the most general aspect of each candidate projection that has the properties that are required will be maintained in the list of candidates. Other aspects will be removed from the list. The favouring of more general aspects may disorientate inexperienced map makers, who will probably expect to obtain a direct aspect as a default. Yet it ensures a maximum reduction of distortion within the area to be mapped, and thus conforms to one of the basic principles underlying the selection procedure (see section 6.3.1).

Ranking of candidate projections

If the list of requested features includes a very restrictive property that is only met by one projection (e.g. straight loxodromes), then the problem of map projection selection is solved. Mostly, however, one will be left with several map projections from which one single projection has to be chosen, and some mechanism for objective ranking will be required. De Genst (1994) presented an objective method for ranking that quantifies the flexibility of a map projection for parameter optimisation (see also De Genst and Canters, 1996). Of all the projections that have the requested properties the one that has the highest flexibility is selected. The method works well for the set of features and projections that was used by De Genst, but it is not generally applicable. We will therefore present a modified version of the method that is based on a few simple decision rules. Since the reduction of overall distortion is a major concern, five rules for ranking have been defined that guarantee that the projections with the most favourable distortion characteristics are retained. These rules are to be applied in the order as given below:

1. If the list of candidates includes different classes of projections priority will be given to the members of the class that has the most general mathematical definition. For the projection classes that are included in the selection procedure proposed in this study the ranking from highest to lowest priority is defined as follows:

    general polyconic projections
    pseudocylindrical projections
    conical projections
    azimuthal / cylindrical projections

2. If the projection is not required to be conformal or equal-area, then priority will be given to those projections that do not have one of both properties.

3. If the pole (or meta-pole) is not to be represented by a point, then priority will be given to projections that represent the pole (or meta-pole) by a line.

Once the first three rules have been applied, the choice will be limited to members of one class of projections, that are either conformal, equal-area or have intermediate distortion characteristics, and that have a pole line or not. The following two rules will gradually reduce the set of candidate projections until the final choice is made:

4. If the set of candidates includes projections with equally spaced (meta-) parallels, or with adjustable spacing of the (meta-)parallels along the central (meta-)meridian, including the possibility for an equal spacing, these projections will be preferred.

5. If the set of candidate projections includes projections with adjustable parameters, the projection with the highest flexibility for graticule optimisation will be selected.

Rule 4 has been included because map projections with equally spaced (meta-) parallels are known to produce well-balanced distortion patterns with less extreme scale variation (see section 2.5.1). Rule 5 guarantees that the optimisation of projection parameters produces a map with the least possible distortion. To determine which of the candidate projections has the most flexible graticule one may simply count the number of independent parameters in the transformation formulas of each projection. Alternatively, one may also give a different weight to each parameter according to its impact on the geometry of the graticule (De Genst and Canters, 1996). If the five above rules do not produce a unique solution, an arbitrary choice will be made.

Which rules will actually be used for discrimination will depend on the class and the characteristics of the projections that are included in the list of candidates. From table 6.4a it is clear that for conical projections the second and third rule unambiguously define the projection that will be chosen. For cylindrical projections, the first two rules determine the projection that is preferred, yet one may still have to choose between one or two standard parallels. Applying rule 5, the version of the projection with two standard parallels will be chosen. For azimuthal projections, the first two rules will produce a unique solution if a conformal or equal-area projection is needed. If none of these two properties is required there may be four candidates left. According to rule 4, the equidistant projection will be selected. For pseudocylindrical and general polyconic projections, which are only used for the mapping of very large areas, graticules with a pole line and equal (or adjustable) spacing of the (meta-)parallels are preferred (rules 3 and 4).

Strictly speaking, the two general polyconic projections that are included in the selection scheme could have been omitted, since all combinations of features that are covered by these projections are also satisfied by their pseudocylindrical counterparts. Polynomial transformation of the latter will automatically produce general polyconic projections with the desired properties. The reason why the two general polyconics have been included anyway, is to offer the map maker the possibility to derive doubly symmetric graticules with curved parallels without the need for polynomial transformation. By choosing for the Wagner projections, a great variety of graticules can be obtained by simply changing or optimising the values of the projection parameters. It is then up to the user to decide if the results of parameter optimisation are satisfactory or not, and if further attempts to reduce distortion are worth the effort.

6.3.4   Implementation issues

The advantages of a dynamic, interactive selection method that couples feature-based selection with optimisation of distortion is demonstrated by the various examples of optimisation that have been shown in chapter 5. Yet the method that is presented can only be fully appreciated if the whole procedure (feature identification and projection optimisation) is integrated in a user-friendly application that guides the user through the different steps of the selection process, and documents intermediate and final results. The information flow diagram that has been presented (figure 6.2) illustrates how the user interface should be designed, dictating the main order of processing and the different feedback mechanisms that should be built in the procedure. A prototype version of a selection tool, which is based on a slightly different set of features and map projections than those that have been presented here, and which uses only part of the decision rules listed in section 6.3.3, has been implemented on Apple Macintosh™ (De Genst, 1994; De Genst and Canters, 1996). The application makes optimal use of the Macintosh toolbox facilities, and follows the standard rules for interface programming defined by Apple (1985, 1992).

For each step in the selection process, appropriate menus are presented to the

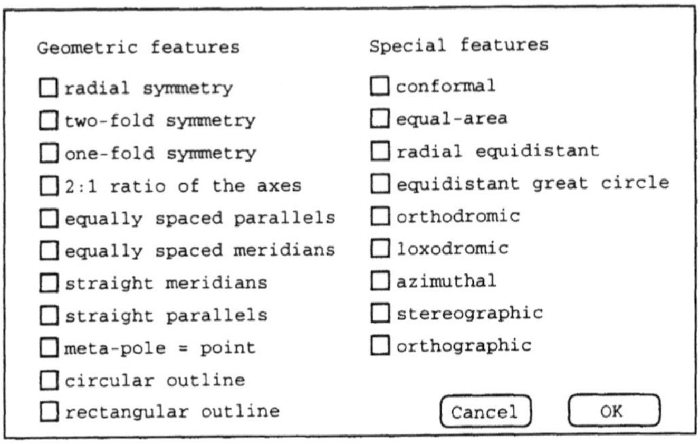

Geometric features	Special features
☐ radial symmetry	☐ conformal
☐ two-fold symmetry	☐ equal-area
☐ one-fold symmetry	☐ radial equidistant
☐ 2:1 ratio of the axes	☐ equidistant great circle
☐ equally spaced parallels	☐ orthodromic
☐ equally spaced meridians	☐ loxodromic
☐ straight meridians	☐ azimuthal
☐ straight parallels	☐ stereographic
☐ meta-pole = point	☐ orthographic
☐ circular outline	
☐ rectangular outline	[Cancel] [OK]

Figure 6.3 The features dialog (De Genst and Canters, 1996).

user in the order dictated by the selection procedure (area – features – aspect) (figure 6.3). The user interface is developed in such a way that the user is unable to select projection properties that are mutually exclusive, or to select aspects that cannot meet the requested properties. To select the aspect of the projection two different menus are used, one for conic projections (three aspects) (figure 6.4) and one for non-conic projections (seven aspects) (figure 6.5). Since the seven aspects of a non-conic projection are not generally known, not even by experienced cartographers, it was decided to use pictograms that show the position of the poles and the equator in each aspect. After area, features and aspect have been defined, the application selects one of the candidate projections by applying rules for ranking as described above. Once the projection has been selected the user may choose to fix the value of some of the projection parameters. In the prototype version of the tool that is presently available, the selection of parameters is restricted to the position of the centre of the projection, the longitude of the central meridian, and the specification of the standard parallels. Other parameters, like the length of the pole line, and the ratio of the axes, which may be important for the design of global representations, cannot be chosen by the user. Dialogues for the choice of free parameters are different for each projection, and also change with the aspect of the projection. If the user specifies a value for each parameter, the projection will be unambiguously defined, and the selection process can be ended. If some of the parameter values are not provided, optimisation will be necessary to obtain a unique solution.

The user may choose between simple parameter optimisation and low-error polynomial transformation. It is suggested that the user first optimises the parameters of the projection, and only applies polynomial transformation methods if the optimisation does not produce a satisfactory result, although this sequence of action is not mandatory. Optimisation is accomplished by generating a set of 5000 distances with a rectangular distance frequency distribution, between 0° and 30° in

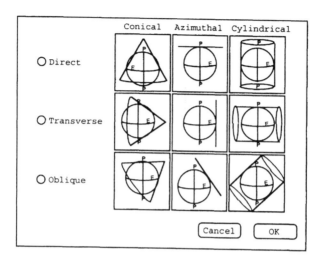

Figure 6.4 The aspect dialog for conic projections (De Genst and Canters, 1996).

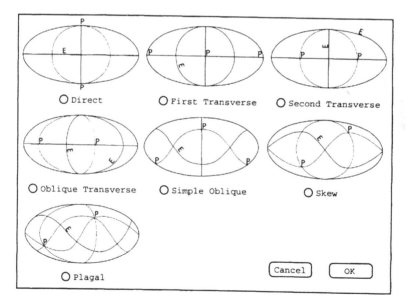

Figure 6.5 The aspect dialog for non-conic projections (De Genst and Canters, 1996).

length (see section 3.6), within a user-defined polygon indicating the region of interest. The mean distortion of these distances is minimised by using the simplex method (see section 5.2). After optimisation, a report with the optimal values for the projection parameters, the polynomial coefficients (in the case of polynomial transformation), and the mean finite scale distortion value is presented. For more detailed information about the implementation, the reader is referred to De Genst and Canters (1996).

It should be pointed out that the selection tool is still in a preliminary state of development. The first part of the selection process (selection of area, features and aspect, choice of projection) has been fully implemented. Yet the programming should be adapted to all the changes that have been made in the selection procedure since the first version of the tool was developed (changes in the list of features and projections, new decision rules). The list of user-adjustable parameters should be extended (see above), and the optimisation process itself should be implemented more efficiently. To allow the user to examine the optimised projection, it should also become possible to show the graticule with lines of equal distortion (isocols) superimposed. At present this can only be done outside the application. It is also investigated how the selection tool could be made available through the web. Although many web sites can be found that are dedicated to map projections, none of them offers advanced facilities for map projection choice or map projection optimisation.

CHAPTER SEVEN

Summary
and conclusions

While the importance of map projection is widely recognised in large scale mapping, its meaning in small-scale map design is less well understood. Most literature about map projection is restricted to a purely mathematical treatment of the subject. In the majority of textbooks on map projection, the selection and use of map projections is given little or no consideration. In consequence even skilled cartographers often have limited notion of the role the projection plays in map design and communication, and do not consider it as important as the other elements in map design. Depending on the map maker's attitude, this may lead to: (a) the repeated use of a projection that for various reasons has become well established for the mapping of a certain area, with no questioning of the suitability of the projection for the map to be constructed; (b) the arbitrary choice of a map projection, without paying any attention to the principles of map projection selection. In both cases there is the risk of misusing a particular projection for a purpose it is not suited for, or at least the risk of not using the projection that would most effectively communicate the map's message.

Present-day map projection software, with its unlimited possibilities for the selection and modification of map projections, merely increases the risk of making improper decisions, especially when used by people that lack a basic understanding of map projections. It is therefore recommended that map projection tools at least offer the user some elementary guidance in choosing a proper map projection. Unfortunately, today this is not the case. The main objective of this study was to propose a strategy for computer-assisted map projection selection that builds upon recent advances in map projection research, and makes optimum use of the possibilities for map projection optimisation offered through digital computing. Research into existing map projection selection schemes indicated that most schemes are either too simple in the definition of their rules, or include only part of the map projection properties that might be considered useful. Also, many schemes do not produce a unique solution because of the large number of existing projections from which a choice must be made. Furthermore, all schemes disregard the fact that the large number of projections that has been presented so far still only represents a fraction of all theoretically possible mappings, and therefore does not necessarily include a satisfying solution for every mapping task. Identification of these shortcomings formed the basis for the research that is described in this monograph.

An important element in the development of the strategy for map projection selection that is proposed, is that no attempt is made to come up with a selection scheme that enables one to select the most appropriate projection from the extensive collection of existing map projections. Since many map projections share the same properties, and may sometimes even hardly differ from one another, this is considered an irrelevant task. Rather than working with a large number of projections, it was decided to define a small set of map projections able to fulfil all combinations of map projection properties that may be useful in small-scale map design. While being able to satisfy each useful combination of features, working with a limited set of map projections cannot guarantee that an optimal distribution of distortion is obtained for the mapping of an arbitrary area, even not if the parameter values for the selected map projection are properly chosen. By applying error-reducing transformations, however, new map projections can be derived from existing ones, with distortion patterns that are optimally adapted to the characteristics of the area to be mapped. This way, maps can be designed with much less overall distortion than those obtained with standard projections. Although not practical in the pre-computer age, the processing power of present-day computers allows for the on-line development of such optimised, low-error projections. Combining map projection selection, based on a small set of map projections, with error-reducing transformation methods, as it is proposed in this study, creates virtually unlimited possibilities for the development of "tailor-made" map projections, especially designed for a particular mapping task. Much more flexibility is obtained than with traditional methods for map projection selection that are based on the direct use of an arbitrary set of standard map projections.

Incorporating methods for the reduction of map projection distortion into a strategy for small-scale map projection selection raises two important questions. First, one has to decide how map projection distortion will be measured. Second, one has to find out what methods for error reduction are most appropriate to be included in the selection process. Both problems have been thoroughly addressed in this study. After a short introduction to the general theory of map projections, and a review of the most important approaches to map projection classification, the measurement and evaluation of map projection distortion was discussed in detail. Special attention was paid to the distortion of large areas, which is an important issue in small-scale mapping. Next, various methods for the modification and transformation of map projections were discussed and evaluated in terms of their suitability for the design of low-error maps with various properties. Methods based on the use of polynomial equations were extensively covered. Since they are the most general of all optimisation methods described in literature, and can easily be adapted to meet user-defined conditions, they are very suited for automated map projection design. Several examples of low-error map projections, obtained through polynomial transformation, are described in this study. Most of these projections have a more favourable distribution of distortion than can be obtained with any standard map projection. In the final chapter a generic method for semi-automated map projection selection is presented, in accordance with the general ideas described above. The method combines rule-based selection with numerical optimisation, and makes extensive use of the different techniques for error-reduction that were discussed in the previous chapters. The most important results of the study, and the main conclusions and recommendations that can be drawn from it will now be discussed in some more detail.

Measurement of map projection distortion

Numerous indices have been proposed to characterise map projection distortion. Some of these indices only reflect a particular aspect of distortion like, for example, the distortion of angles, area or scale. Others have been especially designed to measure the combined effect of different types of distortion. A clear distinction can be made between local indices of distortion, which measure the distortion in one point of the map, using differential geometry, and finite indices, which measure the distortion of finite distances, areas or shapes. Local measurement allows visualisation of the spatial distribution of different types of distortion by way of indicatrices or isocols, and makes it possible to compare the distortion characteristics for various map projections in detail. For an objective comparison of map projection distortion, for the optimisation of map projection parameters, or for the transformation of map projection coordinates to reduce overall error, distortion characteristics for an entire area have to be quantified by one value. Traditionally, this is done by averaging local distortion values for a large number of locations covering the area of the map, using area weight factors. Alternatively, one may also average finite distortion values for a sample of arcs, circles, triangles, or other objects defined on the surface of the sphere.

While distortion indices derived from local measures may be useful to characterise overall distortion for relatively small areas, spatial averaging of local distortion values for a large set of world map projections shows that they are less appropriate to characterise overall distortion for areas of continental size. This can be explained by the fact that indices based on local measurement only represent an average local distortion, and do not take account of the spatial variation of distortion values. It is precisely the spatial variation of distortion that will determine how geographical objects will be displayed on the map, and thus how map projection distortion will be perceived by the map user. Measuring distortion at the finite scale proves to produce more consistent results. Based on a critical review of earlier studies about finite map projection distortion, new indices for measuring the overall distortion of scale, area and shape are presented. The proposed measure for finite scale distortion, which compares distances on the map to corresponding distances on the globe for a sample of great-circle arcs covering the area to be mapped, optimally accounts for the spatial variation in local scale distortion. Calculating overall finite scale distortion for the continental area for a selection of world map projections demonstrates the measure's sensitivity to changes in the geometry of the graticule. Since map projections with a low finite scale distortion prove to have moderate distortion of area and shape, the overall distortion of scale, as it is defined in this study, is suitable as an optimisation criterion, not only for the development of low-error conformal and equal-area projections, but also for the design of so-called intermediate projections. Intermediate projections are neither conformal nor equal-area, but do not have the large distortions of area and angles, characteristic of conformal and equal-area graticules. Due to the more balanced patterns of distortion of these projections, a much better representation of large areas can be obtained. For that reason, intermediate map projections are very popular for global mapping. Several examples of low-error intermediate projections, especially designed for world maps, are presented in this study.

Modification and transformation of map projections

Once a proper measure for overall map projection distortion has been agreed upon, one still has to define how error reduction will be achieved. What methods for error reduction are most appropriate to be included in a procedure for map projection selection depends on the generality of these methods, their ability to reduce map projection distortion, and the possibilities they offer to impose various map purpose related constraints. Indeed, since selection starts with the definition of a number of geometric and/or special properties the map projection should have, it is required that these properties are maintained in the optimisation process. A reduction of map projection distortion can be accomplished in two ways. One may start from a set of general map projection equations, and optimise the value of one or more of the parameters in these equations. Requested properties may be obtained by putting appropriate constraints on the values of the parameters. This process is referred to as *modification*. Alternatively, one may start with a particular map projection, and transform the map projection coordinates by applying a planar transformation that maintains some of the properties of the original graticule. This process is known as map projection *transformation*. In this study, various types of map projection modification and transformation have been discussed and applied, to see how effective they are in reducing map projection distortion, and to find out if they might be considered for inclusion in a tool for automated map projection selection and optimisation, or not.

A very simple, well-known type of modification is the optimal positioning of the meta-pole and the line(s) of zero distortion for a conic projection. Adjustment of aspect and projection parameters for conic type projections can be seen as the most elementary form of map projection optimisation, and should be available in any tool for automated map projection selection. While the oblique aspect of the azimuthal projection is well-known, oblique aspects of cylindrical and conical projections are rarely applied. As such, the possibilities for error reduction offered by standard map projections of the conic type are largely underestimated. Optimisation of the projection parameters for a map of North and South America (see section 5.4.1) proves that conic type projections can be used with satisfaction for the mapping of extended areas, at least if the aspect of the projection can be chosen freely, and if avoidance of visible distortion is the major concern. Only if additional geometric conditions are imposed, or if higher standards of geometric accuracy have to be met, or if the area to be mapped is extremely large (as for world maps), will the use of other map projections, and/or more complicated optimisation methods be required.

Very interesting, especially for the derivation of optimal projections for world maps, are Wagner's generalised pseudocylindrical and polyconic map projection equations, which allow independent adjustment of various graticule characteristics through a proper choice of parameter values. Optimisation of the values of the parameters for the entire continental area, not taking account of specific geometric conditions, produces low-error equal-area and intermediate projections with distortion patterns that are adapted to the distribution of the continents, and with overall distortion values that are lower than for any standard map projection of the corresponding type. Unfortunately, Wagner's generalised map projection equations do not have the level of generality that is needed to be able to develop low-error map projections that satisfy different geometric conditions. Adjusting the values of the parameters in Wagner's formulae produces graticules with meridians and parallels of the same type as on the standard map projections from which the generalised equations are derived. Once one or two geometric conditions are imposed there is little or no freedom left for further optimisation.

A more general approach to low-error map projection design consists of expressing the relationship between map projection coordinates and geographical longitude and latitude by means of general bivariate polynomials, and optimising the values of the polynomial coefficients. Using polynomial map projection equations, a multitude of new graticules can be derived, varying from graticules with a very complex, irregular geometry, to graticules that belong to traditional map projection classes. As has been shown in this study, many geometric properties that are useful in small-scale mapping can be imposed by putting some of the polynomial coefficients in the general mapping equations equal to zero, others by defining a simple mathematical relationship between two or more coefficients. Only for the development of low-error map projections that represent the pole by a point, a slightly different equation format has to be used. Starting from general polynomial equations, a whole set of low-error polynomial projections with intermediate distortion characteristics, suitable for world maps, has been presented, each one with its own particular geometric properties. All of these graticules have less overall scale distortion than any standard map projection with the same properties, presently used for global mapping.

Instead of using polynomial equations to transform positions from the curved surface of the Earth directly onto the map, low-error map projections can also be obtained by transforming the coordinates of any standard map projection in the map plane, again using polynomial type equations. By optimising the values of the coefficients that define the transformation, the patterns of distortion of the original projection will be adapted to the shape of the area to be mapped, which will result in less overall distortion. An advantage of this method is that not only the geometric properties of the original projection, but also the equal-area property and the correct representation of angles can easily be maintained by controlling the coefficient values of the polynomial transformation. Although various authors have used complex-algebra polynomials to transform standard conformal map projections into conformal projections with less distortion for a selected area, the use of polynomial transformation for the development of low-error intermediate or equal-area projections, both important for small-scale mapping, has hardly been explored. In this study attention has been focused on these two types of transformation.

For the derivation of low-error intermediate projections, general bivariate polynomials identical to the ones applied for direct transformation can be used. Only the geographical latitude and longitude in the equations have to be replaced by the x and y map coordinates of the original projection. Formulation of geometric conditions is done in exactly the same way as for the direct case. One difference, however, is that the same set of equations can be used for the development of projections with and without a pole line. Whether the pole is represented by a point or a line will be determined by the projection that is transformed. For the design of low-error equal-area projections, two polynomial transformations have been proposed that can be combined to one double transformation with a high level of flexibility. Geometric constraints are again easy to impose. Various examples of low-error equal-area projections obtained with the new transformation have been presented, for use in global, continental, as well as regional mapping. The proposed map projections, with lines of equal distortion closely adapted to the shape of the area to be mapped, can be considered as equal-area counterparts of the better known low-error conformal projections derived through complex-algebra polynomial transformation. The examples include a set of low-error equal-area projections for the European Union which all have much less scale distortion than can be obtained with any standard equal-area projection.

Semi-automated map projection selection

Based on a critical review of previously proposed schemes, seven general principles that should be taken into account when developing a computer-based method for small-scale map projection selection were identified:

1. Map projection selection should be guided by the purpose of the map.
2. The method should produce a unique solution for any application.
3. The selected projection must have low overall distortion for the area to be mapped.
4. The method must be applicable to global, continental and regional mapping.
5. The method should work for each set of map projection features.
6. The map maker should be able to evaluate and possibly reject the selected projection.
7. The method should take full advantage of available processing power and possibilities for computer-based interaction.

To be able to fulfil these conditions, it was concluded that selection should best be based on a small set of standard map projections that includes at least one projection for any combination of map projection properties that might be considered useful in small-scale mapping. Since even with a small set of projections certain combinations of properties may lead to the selection of more than one candidate projection a simple set of rules was proposed to rank all candidate projections. The ranking favours those projections that have the most balanced distortion patterns and the highest degree of freedom for adapting the graticule's geometry to the area to be mapped. Once the most suitable projection has been identified, numerical optimisation can be applied to define the most appropriate values for the parameters of the projection, and possibly, to transform the projection using polynomial type equations, to obtain a low-error solution that is optimally adapted to the mapping area.

The method that is proposed thus consists of two consecutive processes: (1) the selection of a map projection that has all the properties requested by the map maker, and (2) the optimisation of the projection's graticule to minimise distortion within the area to be mapped. To be applicable on all scale levels it is suggested that the list of candidate projections is derived in three steps, using the following criteria: (a) the map projection features that are required; (b) the extension of the area to be mapped; (c) the aspect of the projection. The main reason for having to specify the extension of the area to be mapped is that very different types of map projections are used for global, hemispherical and continental (or regional) mapping. Specification of the aspect is not strictly required, but should be available as an option. Since the most general of all aspects (the plagal aspect) will ensure a maximum reduction of distortion in the optimisation phase, and will therefore be selected by default, the map maker should be able to select less complicated aspects, if the purpose of the map dictates a simple, more straightforward graticule geometry.

A very important element in the selection method that is presented is the interaction with the map maker. Numerous examples of low-error maps shown in this study have made it clear that optimisation of projection parameters, or transformation of map projections to reduce overall distortion, not necessarily produces acceptable maps. Some map projection properties, imposed by the map maker, may make it difficult, if not impossible to obtain a good spatial distribution of distortion, while other properties, which may not have been proposed at the beginning, may contribute to a more balanced pattern of distortion, with lower distortion values at the extremities of the area. Map projection selection, therefore, should be seen as an essentially dynamic process, where map projection properties are repeatedly updated by the map maker until an optimal result is obtained. This implies that the map maker must be able to properly evaluate the outcome of the optimisation process and to indicate what actions should be taken for improvement. Detailed error reports and maps showing the distribution of various aspects of map projection distortion should be made available. Map projection specifications should be editable at all times.

From the foregoing, it is clear that the proposed selection method still requires a great deal of human intervention. The use of optimisation techniques, as suggested in this study, can only help the map maker to reach an appropriate solution and increase the level of objectivity in the decision process. However, it is the map maker who ultimately decides what properties the map projection should have, and if the proposed solution is satisfactory or not. This implies that at least some basic knowledge of map projection principles is expected. Alternatively, one might think of developing a selection system that is able to identify map projection properties directly from the type of application. This was not pursued in this study. The relationship between map function and map use on the one hand, and map projection properties on the other, is not well understood, and is very difficult to formalise. Very often the selection process involves various trade-offs that cannot be translated into simple rules. Combining map projection selection with advanced methods for optimisation further complicates the proper choice of map projection properties. It is therefore doubtful if it will ever be possible to develop a selection system, which does not require some knowledge on the part of the map maker, unless it concerns a simple method with limited options for map projection improvement that produces sub-optimal solutions, or that is only applicable within a particular range of scales.

APPENDIX A

Features for the non-direct aspects of all map projections used in the selection procedure

This appendix presents seven tables describing useful features for the non-direct aspects of all map projections included in the selection procedure proposed in chapter six. Tables A.1 and A.2 list all features for the transverse and oblique aspect of the conic-type map projections, while tables A.3 through A.7 list all features for the six non-direct aspects of the non-conic map projections used in the selection.

Table A.1 Features for the transverse aspect of all conic map projections included in the selection procedure.

MAP PROJECTIONS	Radial symmetry	Double symmetry	Single symmetry	Equal spacing along the central meridian	Equal spacing along the equator	Straight parallels	Straight meridians	Circular outline	Rectangular outline	Oval outline	Meta-pole is represented by a point	2:1 ratio of the axes	Conformal	Equal-area	Radial equidistant	Equidistant along the meta-equator	Azimuthal	Loxodromic	Orthodromic	Stereographic	Orthographic	Perspective
Azimuthal																						
Azimuthal equidistant		x	x	x	x			x			x				x	x	x					
Azimuthal equal-area		x	x					x			x	x		x			x					
Azimuthal stereographic		x	x					x			x		x				x			x		
Azimuthal gnomonic		x	x			x	x	x			x						x		x			
Azimuthal orthographic		x	x					x			x						x				x	
Azimuthal outer-perspective		x	x					x			x						x					x
Cylindrical																						
Cylindrical equidistant one standard parallel		x	x	x	x				x							x						
Cylindrical equal-area one standard parallel		x	x	x	x				x					x								
Cylindrical conformal one standard parallel		x	x	x					x				x									
Conical																						
Conical equidistant			x	x	x				x							x						
Conical equal-area			x	x					x					x								
Conical conformal			x	x					x		x		x									

Table A.2 Features for the oblique aspect of all conic map projections included in the selection procedure.

MAP PROJECTIONS	Radial symmetry	Double symmetry	Single symmetry	Equal spacing along the central meridian	Equal spacing along the equator	Straight parallels	Straight meridians	Circular outline	Rectangular outline	Oval outline	Meta-pole is represented by a point	2:1 ratio of the axes	Conformal	Equal-area	Radial equidistant	Equidistant along the meta-equator	Azimuthal	Loxodromic	Orthodromic	Stereographic	Orthographic	Perspective
Azimuthal																						
Azimuthal equidistant			×	×				×			×				×		×					
Azimuthal equal-area			×					×			×			×			×					
Azimuthal stereographic			×					×			×		×				×			×		
Azimuthal gnomonic			×				×	×			×						×		×			
Azimuthal orthographic			×					×			×						×				×	
Azimuthal outer-perspective			×					×			×						×					×
Cylindrical																						
Cylindrical equidistant one standard parallel			×	×					×							×						
Cylindrical equal-area one standard parallel			×	×					×			×		×		×						
Cylindrical conformal one standard parallel			×						×				×			×						
Conical																						
Conical equidistant			×						×													
Conical equal-area			×						×					×								
Conical conformal			×						×				×									

Table A.3 Features for the first transverse aspect of all non-conic map projections included in the selection procedure.

MAP PROJECTIONS	Radial symmetry	Double symmetry	Single symmetry	Equal spacing along the central meridian	Equal spacing along the equator	Straight parallels	Straight meridians	Circular outline	Rectangular outline	Oval outline	Meta-pole is represented by a point	2:1 ratio of the axes	Conformal	Equal-area	Radial equidistant	Equidistant along the meta-equator	Azimuthal	Loxodromic	Orthodromic	Stereographic	Orthographic	Perspective
Pseudocylindrical																						
Wagner VI																						
Equidistant meta-equator		×	×	×												×						
2:1 ratio of the axes		×	×	×								×										
Equidistant meta-equator + 2:1 ratio of the axes		×	×	×								×				×						
Metapole is mapped as a point equidistant meta-equator		×	×	×						×	×					×						
2:1 ratio of the axes		×	×	×						×	×	×										
equidistant meta-equator + 2:1 ratio of the axes		×	×	×						×	×	×				×						
Wagner I																						
Equidistant meta-equator		×	×	×										×		×						
2:1 ratio of the axes		×	×	×								×		×								
Metapole is mapped as a point equidistant meta-equator + 2:1 ratio of the axes		×	×	×						×	×	×		×		×						
Mollweide		×	×	×						×	×	×		×		×						

Table A.3 continued.

MAP PROJECTIONS	Radial symmetry	Double symmetry	Single symmetry	Equal spacing along the central meridian	Equal spacing along the equator	Straight parallels	Straight meridians	Circular outline	Rectangular outline	Oval outline	Meta-pole is represented by a point	2:1 ratio of the axes	Conformal	Equal-area	Radial equidistant	Equidistant along the meta-equator	Azimuthal	Loxodromic	Orthodromic	Stereographic	Orthographic	Perspective
Polyconic																						
Aitoff-Wagner																						
Equidistant meta-equator		x	x	x												x						
2:1 ratio of the axes		x	x	x								x										
Equidistant meta-equator + 2:1 ratio of the axes		x	x	x								x				x						
Metapole is mapped as a point		x	x	x							x											
equidistant meta-equator		x	x	x							x					x						
2:1 ratio of the axes		x	x	x							x	x										
equidistant meta-equator + 2:1 ratio of the axes		x	x	x							x	x				x						
oval outline		x	x	x						x	x											
equidistant meta-equator		x	x	x						x	x					x						
2:1 ratio of the axes		x	x	x						x	x	x										
equidistant meta-equator + 2:1 ratio of the axes		x	x	x						x	x	x				x						
Hammer-Wagner																						
2:1 ratio of the axes		x	x									x		x								
Metapole is mapped as a point		x	x								x			x								
2:1 ratio of the axes		x	x								x	x		x								
oval outline		x	x							x	x			x								
2:1 ratio of the axes		x	x							x	x	x		x								

Table A.4 Features for the second transverse aspect of all non-conic map projections included in the selection procedure.

MAP PROJECTIONS	GEOMETRIC FEATURES												SPECIAL FEATURES									
	Graticule							Meta-graticule					Locally defined					Globally defined				
	Symmetry			Spacing/shape				Outline							Equidistant								
	Radial symmetry	Double symmetry	Single symmetry	Equal spacing along the central meridian	Equal spacing along the equator	Straight parallels	Straight meridians	Circular outline	Rectangular outline	Oval outline	Meta-pole is represented by a point	2:1 ratio of the axes	Conformal	Equal-area	Radial equidistant	Equidistant along the meta-equator	Azimuthal	Loxodromic	Orthodromic	Stereographic	Orthographic	Perspective
Pseudocylindrical																						
Wagner VI																						
Equidistant meta-equator		×	×	×	×											×						
2:1 ratio of the axes		×	×	×	×							×				×						
Equidistant meta-equator + 2:1 ratio of the axes		×	×	×	×							×				×						
Metapole is mapped as a point																						
equidistant meta-equator		×	×	×	×					×	×					×						
2:1 ratio of the axes		×	×	×	×					×	×	×										
equidistant meta-equator + 2:1 ratio of the axes		×	×	×	×					×	×	×				×						
Wagner I																						
Equidistant meta-equator		×	×	×	×									×		×						
2:1 ratio of the axes		×	×	×	×							×		×								
Metapole is mapped as a point																						
equidistant meta-equator + 2:1 ratio of the axes		×	×	×						×	×	×		×								
Mollweide		×	×	×						×				×		×						

Table A.4 continued.

MAP PROJECTIONS	GEOMETRIC FEATURES												SPECIAL FEATURES									
	Graticule							Meta-graticule					Locally defined				Globally defined					
	Symmetry			Spacing/shape				Outline					Conformal	Equal-area	Radial equidistant	Equidistant along the meta-equator	Azimuthal	Loxodromic	Orthodromic	Stereographic	Orthographic	Perspective
	Radial symmetry	Double symmetry	Single symmetry	Equal spacing along the central meridian	Equal spacing along the equator	Straight parallels	Straight meridians	Circular outline	Rectangular outline	Oval outline	Meta-pole is represented by a point	2:1 ratio of the axes	Conformal	Equal-area	Radial equidistant	Equidistant along the meta-equator	Azimuthal	Loxodromic	Orthodromic	Stereographic	Orthographic	Perspective
Polyconic																						
Aitoff-Wagner																						
Equidistant meta-equator		x	x	x	x											x						
2:1 ratio of the axes		x	x	x	x							x										
Equidistant meta-equator + 2:1 ratio of the axes		x	x	x	x							x				x						
Metapole is mapped as a point																						
equidistant meta-equator		x	x	x	x						x					x						
2:1 ratio of the axes		x	x	x	x						x	x										
equidistant meta-equator + 2:1 ratio of the axes		x	x	x	x						x	x				x						
oval outline																						
equidistant meta-equator		x	x	x	x					x	x					x						
2:1 ratio of the axes		x	x	x	x					x	x	x										
equidistant meta-equator + 2:1 ratio of the axes		x	x	x	x					x	x	x				x						
Hammer-Wagner																						
2:1 ratio of the axes		x	x	x	x							x		x								
Metapole is mapped as a point																						
2:1 ratio of the axes		x	x								x	x		x								
oval outline																						
2:1 ratio of the axes		x	x							x	x	x		x								

Table A.5 Features for the transverse oblique aspect of all non-conic map projections included in the selection procedure.

MAP PROJECTIONS	Radial symmetry	Double symmetry	Single symmetry	Equal spacing along the central meridian	Equal spacing along the equator	Straight parallels	Straight meridians	Circular outline	Rectangular outline	Oval outline	Meta-pole is represented by a point	2:1 ratio of the axes	Conformal	Equal-area	Radial equidistant	Equidistant along the meta-equator	Azimuthal	Loxodromic	Orthodromic	Stereographic	Orthographic	Perspective
Pseudocylindrical																						
Wagner VI																						
Equidistant meta-equator			x	x												x						
2:1 ratio of the axes			x	x								x										
Equidistant meta-equator + 2:1 ratio of the axes			x	x								x				x						
Metapole is mapped as a point																						
equidistant meta-equator			x	x						x	x					x						
2:1 ratio of the axes			x	x						x	x	x										
equidistant meta-equator + 2:1 ratio of the axes			x	x						x	x	x				x						
Wagner I																						
Equidistant meta-equator			x	x										x		x						
2:1 ratio of the axes			x	x								x		x								
Metapole is mapped as a point																						
equidistant meta-equator + 2:1 ratio of the axes			x	x						x	x	x		x		x						
Mollweide			x	x						x	x	x		x								

Table A.5 continued.

MAP PROJECTIONS	Radial symmetry	Double symmetry	Single symmetry	Equal spacing along the central meridian	Equal spacing along the equator	Straight parallels	Straight meridians	Circular outline	Rectangular outline	Oval outline	Meta-pole is represented by a point	2:1 ratio of the axes	Conformal	Equal-area	Radial equidistant	Equidistant along the meta-equator	Azimuthal	Loxodromic	Orthodromic	Stereographic	Orthographic	Perspective
Polyconic																						
Aitoff-Wagner																						
Equidistant meta-equator			×	×												×						
2:1 ratio of the axes			×	×								×										
Equidistant meta-equator + 2:1 ratio of the axes			×	×								×				×						
Metapole is mapped as a point											×											
equidistant meta-equator			×	×							×					×						
2:1 ratio of the axes			×	×							×	×										
equidistant meta-equator + 2:1 ratio of the axes			×	×							×	×				×						
oval outline										×	×											
equidistant meta-equator			×	×						×	×					×						
2:1 ratio of the axes			×	×						×	×	×										
equidistant meta-equator + 2:1 ratio of the axes			×	×						×	×	×				×						
Hammer-Wagner																						
2:1 ratio of the axes			×	×								×		×								
Metapole is mapped as a point											×			×								
2:1 ratio of the axes			×	×							×	×		×								
oval outline										×	×			×								
2:1 ratio of the axes			×	×						×	×	×		×								

Table A.6 Features for the simple oblique aspect of all non-conic map projections included in the selection procedure.

MAP PROJECTIONS	Radial symmetry	Double symmetry	Single symmetry	Equal spacing along the central meridian	Equal spacing along the equator	Straight parallels	Straight meridians	Circular outline	Rectangular outline	Oval outline	Meta-pole is represented by a point	2:1 ratio of the axes	Conformal	Equal-area	Radial equidistant	Equidistant along the meta-equator	Azimuthal	Loxodromic	Orthodromic	Stereographic	Orthographic	Perspective
Pseudocylindrical																						
Wagner VI																						
Equidistant meta-equator			x	x												x						
2:1 ratio of the axes			x	x								x										
Equidistant meta-equator + 2:1 ratio of the axes			x	x								x				x						
Metapole is mapped as a point																						
equidistant meta-equator			x	x						x	x			x		x						
2:1 ratio of the axes			x	x						x	x	x		x								
equidistant meta-equator + 2:1 ratio of the axes			x	x						x	x	x		x		x						
Wagner I																						
Equidistant meta-equator			x	x										x		x						
2:1 ratio of the axes			x	x								x		x								
Metapole is mapped as a point																						
equidistant meta-equator + 2:1 ratio of the axes			x							x	x	x		x		x						
Mollweide			x							x	x	x		x		x						

Table A.6 continued.

Column groups — GEOMETRIC FEATURES: *Graticule* → Symmetry (Radial symmetry, Double symmetry, Single symmetry); Spacing/shape (Equal spacing along the central meridian, Equal spacing along the equator, Straight parallels, Straight meridians). *Meta-graticule* → Outline (Circular outline, Rectangular outline, Oval outline); Meta-pole is represented by a point; 2:1 ratio of the axes. SPECIAL FEATURES: *Locally defined* → Conformal, Equal-area, Equidistant (Radial equidistant, Equidistant along the meta-equator). *Globally defined* → Azimuthal, Loxodromic, Orthodromic, Stereographic, Orthographic, Perspective.

MAP PROJECTIONS	Radial symmetry	Double symmetry	Single symmetry	Equal spacing along the central meridian	Equal spacing along the equator	Straight parallels	Straight meridians	Circular outline	Rectangular outline	Oval outline	Meta-pole is represented by a point	2:1 ratio of the axes	Conformal	Equal-area	Radial equidistant	Equidistant along the meta-equator	Azimuthal	Loxodromic	Orthodromic	Stereographic	Orthographic	Perspective
Polyconic																						
Aitoff-Wagner																						
Equidistant meta-equator			×	×												×						
2:1 ratio of the axes			×	×								×										
Equidistant meta-equator + 2:1 ratio of the axes			×	×								×				×						
Metapole is mapped as a point			×	×							×											
equidistant meta-equator			×	×							×					×						
2:1 ratio of the axes			×	×							×	×										
equidistant meta-equator + 2:1 ratio of the axes			×	×							×	×				×						
oval outline			×	×						×	×											
equidistant meta-equator			×	×						×	×			×		×						
2:1 ratio of the axes			×	×						×	×	×		×								
equidistant meta-equator + 2:1 ratio of the axes			×	×						×	×	×		×		×						
Hammer-Wagner																						
2:1 ratio of the axes			×								×	×		×								
Metapole is mapped as a point			×								×			×								
oval outline			×							×	×			×								
2:1 ratio of the axes			×							×	×	×		×								

308

Table A.7 Features for the skew and plagal aspects of all non-conic map projections included in the selection procedure.

MAP PROJECTIONS	Radial symmetry	Double symmetry	Single symmetry	Equal spacing along the central meridian	Equal spacing along the equator	Straight parallels	Straight meridians	Circular outline	Rectangular outline	Oval outline	Meta-pole is represented by a point	2:1 ratio of the axes	Conformal	Equal-area	Radial equidistant	Equidistant along the meta-equator	Azimuthal	Loxodromic	Orthodromic	Stereographic	Orthographic	Perspective
Pseudocylindrical																						
Wagner VI																						
Equidistant meta-equator										x						x						
2:1 ratio of the axes										x		x										
Equidistant meta-equator + 2:1 ratio of the axes										x		x				x						
Metapole is mapped as a point																						
equidistant meta-equator										x	x					x						
2:1 ratio of the axes											x	x										
equidistant meta-equator + 2:1 ratio of the axes											x	x				x						
Wagner I																						
Equidistant meta-equator														x		x						
2:1 ratio of the axes												x		x								
Metapole is mapped as a point											x			x								
equidistant meta-equator + 2:1 ratio of the axes											x	x		x		x						
Mollweide										x	x	x		x								

Table A.7 continued.

MAP PROJECTIONS	Radial symmetry	Double symmetry	Single symmetry	Equal spacing along the central meridian	Equal spacing along the equator	Straight parallels	Straight meridians	Circular outline	Rectangular outline	Oval outline	Meta-pole is represented by a point	2:1 ratio of the axes	Conformal	Equal-area	Radial equidistant	Equidistant along the meta-equator	Azimuthal	Loxodromic	Orthodromic	Stereographic	Orthographic	Perspective
Polyconic																						
Aitoff-Wagner																						
Equidistant meta-equator																x						
2:1 ratio of the axes												x										
Equidistant meta-equator + 2:1 ratio of the axes												x				x						
Metapole is mapped as a point											x											
equidistant meta-equator											x					x						
2:1 ratio of the axes											x	x										
equidistant meta-equator + 2:1 ratio of the axes											x	x				x						
oval outline										x	x											
equidistant meta-equator										x	x					x						
2:1 ratio of the axes										x	x	x										
equidistant meta-equator + 2:1 ratio of the axes										x	x	x				x						
Hammer-Wagner														x								
2:1 ratio of the axes												x		x								
Metapole is mapped as a point											x			x								
2:1 ratio of the axes											x	x		x								
oval outline										x	x			x								
2:1 ratio of the axes										x	x	x		x								

References

Airy, G.B., 1861, Explanation of a projection by balance of errors for maps applying to a very large extent of the Earth's surface, and comparison of this projection with other projections, *London, Edinburgh and Dublin Philosophical Magazine*, 4th ser., **22**, 409–21.

Aitoff, D., 1889, Projections des cartes géographiques, in *Atlas de Géographie Moderne*, Paris: Hachette.

Albinus, H.-J., 1979, Lokale und globale Aspekte bei Verzerrungsfunktionen kartographischer Netzentwürfe, *Kartographische Nachrichten*, **29**, 90–3.

Albinus, H.-J., 1981, Anmerkungen und Kritik zur Entfernungsverzerrung, *Kartographische Nachrichten*, **31**, 179–83.

American Cartographic Association, 1989, The case against rectangular world maps, *Cartographic Journal*, **26**, 156–7.

American Cartographic Association (Ed.), 1991, *Matching the Map Projection to the Need*, Bethesda: American Congress on Surveying and Mapping.

Apple, 1985, *Inside Macintosh*, Vol. I, II and III, New York: Addison-Wesley.

Apple, 1992, *Inside Macintosh: Macintosh Toolbox Essentials*, New York: Addison-Wesley.

Arden-Close, C.F., 1938, A polar-azimuthal retroazimuthal projection, *Geographical Journal*, **92**, 536–7.

Arden-Close, C.F., 1952, A forgotten pseudo-zenithal projection, *Geographical Journal*, **118**, 237.

Baar, E.J., 1947, The manipulation of projections for world maps, *Geographical Review*, **37**, 112–20.

Baetslé, P.L., 1970, Optimalisation d'une representation cartographique, *Bulletin trimestriel de la Société Belge de Photogrammetrie*, 11–26.

Baker, J.G.P., 1986, The "Dinomic" world map projection, *Cartographic Journal*, **23**, 66–7.

Baranyi, J., 1968, The problems of the representation of the globe on a plane with special reference to the preservation of the forms of continents, *Hungarian Cartographic Studies*, Budapest: Földmérési Intézet, pp. 19–43.

Baranyi, J. and Györffy, J., 1989, New form-true projections in Hungarian atlases, *Hungarian Cartographical Studies*, Budapest: Földmérési Intézet, pp. 75–85.

Bartholomew, J., 1942, *The Citizen's Atlas of the World*, Edinburgh: John Bartholomew & Son.

Bartholomew, J., 1948, *The Regional Atlas of the World*, Edinburgh: Geographical Institute.

Bartholomew, J., 1958, *The Times Atlas of the World*, Vol. 1, London: Times Publishing.

Behrmann, W., 1909, Zur Kritik der flächentreuen Projektionen der ganzen Erde und einer Halbkugel, *Sitzungsberichte der Königlich Bayerischen Akademie der Wissenschaften*, Mathematisch-physikalische Klasse, München, **13**, 19–48.

Blair, D.J. and Biss, T.H., 1967, *The Measurement of Shape in Geography: An Appraisal of Methods and Techniques*, Bulletin of Quantitative Data for Geographers, no. 11, Department of Geography, Nottingham University.

Bludau, A., 1891, Die flächentreue transversale Kegel-Projektion für die Karte von Afrika, *Zeitschrift der Gesellschaft für Erdkunde*, Berlin, **26**, 145–58.

Bludau, A., 1892, Die flächentreue Azimut-Projektion Lamberts für die Karte von Afrika, *Petermanns Mitteilungen*, **38**, 214–18.

Boggs, S.W., 1929, A new equal-area projection for world maps, *Geographical Journal*, **73**, 241–5.

Boginskiy, V.M., 1972, *Sposob Izyskaniya Proizvol'nykh Proyektsiy Melkomasshtabnykh Kart*, Moscow: Nedra.

Boyce, R.R. and Clark, W.A.V., 1964, The concept of shape in geography, *Geographical Review*, **54**, 561–72.

Bretterbauer, K., 1989, Die trimetrische Projektion von W. Chamberlin, *Kartographische Nachrichten*, **39**, 51–5.

Briesemeister, W., 1953, A new oblique equal-area projection, *Geographical Review*, **43**, 260–1.

Briggs, D.J. and Mounsey, H.M., 1989, Integrating land resource data into a European geographical information system, *Applied Geography*, **9**, 5–20.

Brinker, D.M., 1990, Return of the orthographic projection as the best choice for world maps, *ACSM–ASPRS Annual Convention, Technical Papers, vol. 2: Cartography*, Bethesda, 60–7.

Bugayevskiy, L.M., 1982, Kriterii otsenki pri vybore kartograficheskikh proyektsii, *Geodeziya i Aerofotos'emka*, **3**, 92–6.

Bugayevskiy, L.M. and Snyder, J.P., 1995, *Map Projections, A Reference Manual*, London: Taylor & Francis.

Bunge, W., 1966, *Theoretical Geography*, Lund Studies in Geography, Ser. C, General and Mathematical Geography, 1, Department of Geography, The Royal University of Lund, Sweden.

Buttenfield, B.P. and McMaster, R.B. (Eds.), 1991, *Map Generalization, Making Rules for Knowledge Representation*, Harlow: Longman.

Canters, F., 1989, New projections for world maps, a quantitative-perceptive approach, *Cartographica*, **26**, 53–71.

Canters, F., 1991, Map projection in a GIS environment, in Rybaczuk, K. and Blakemore, M. (Eds.), *Mapping the Nations, Proceedings of the 15th Conference of the International Cartographic Association*, Bournemouth, Vol. 2, pp. 595–604.

Canters, F., 1992, The transformation of positional data for GIS applications, *Proceedings of the Third European Conference and Exhibition on Geographical Information Systems*, Munich, pp. 732–41.

Canters, F., 1995, A dynamic procedure for the selection of an appropriate GIS projection framework, in *Space Scientific Research in Belgium, Vol. III, Earth*

Observation, Part 2, Brussels: Federal Office for Scientific, Technical and Cultural Affairs, pp. 77–90.

Canters, F. and Decleir, H., 1989, *The World in Perspective, A Directory of World Map Projections*, Chichester: John Wiley & Sons.

Canters, F. and De Genst, W., 1997, Development of a low-error equal-area map projection for the European Union, *Proceedings of the 18th ICA/ACI International Cartographic Conference*, Stockholm, pp. 775–82.

Castner, H.W., 1991, Viewing the Earth from space, in American Cartographic Association (Ed.), *Matching the Map Projection to the Need*, Bethesda: American Congress on Surveying and Mapping, pp. 14–15.

Cauvin, C. and Schneider, C., 1989, Cartographic transformations and the piezopleth maps method, *Cartographic Journal*, **26**, 96–104.

Chamberlin, W., 1947, *The Round Earth on Flat Paper: Map Projections Used by Cartographers*, Washington: National Geographic Society.

Chebyshev, P.L., 1856, Sur la construction des cartes géographiques, *Bulletin de l'Académie Impériale des Sciences, Classe Physico-Mathématique*, **14**, 257–61.

Christensen, A.H.J., 1992, The Chamberlin trimetric projection, *Cartography and Geographic Information Systems*, **19**, 88–100.

Clark, D.M., Hastings, D.A. and Kineman, J.J., 1991, Global databases and their implications for GIS, in Maguire, D.J., Goodchild, M.F. and Rhind, D.W. (Eds.), *Geographical Information Systems, Vol. 2: Applications*, Harlow: Longman, pp. 217–31.

Clark, J.W., 1977, Time-distance transformations of transportation networks, *Geographical Analysis*, **9**, 195–206.

Clarke, A.R., 1879, Geography - mathematical geography, in *Encyclopaedia Britannica*, 9th ed., Vol. 10, pp. 197–210.

Close, C.F., 1921, Note on a doubly-equidistant projection, *Geographical Journal*, **57**, 446–8.

Close, C.F., 1922, Note on two double, or two-point, map projections, in *Ordnance Survey, Professional Papers*, new ser., no. 5, London: Ordnance Survey.

Craig, J.I., 1910, *Map Projections, The Theory of Map Projections, with Special Reference to the Projections Used in the Survey Department*, paper no. 13, Cairo: Ministry of Finance, Survey Department.

Craster, J.E.E., 1929, Some equal-area projections of the sphere, *Geographical Journal*, **74**, 471–4.

Dahlberg, R.E., 1962, Evolution of interrupted map projections, *International Yearbook of Cartography*, **2**, 36–54.

Dahlberg, R.E, 1991, Shaping the world map, in American Cartographic Association (Ed.), *Matching the Map Projection to the Need*, Bethesda: American Congress on Surveying and Mapping, pp. 6–7.

Deakin, R.E., 1990, The "tilted camera" perspective projection of the Earth, *Cartographic Journal*, **27**, 7–14.

Decleir, H. and Canters, F., 1986, Projektiesystemen in de atlaskartografie van de 16de en 17de eeuw, in: *Oude kaarten en plattegronden, Bronnen voor de historische geografie van de Zuidelijke Nederlanden*, Archief- en Bibliotheekwezen in België, extra nr. 31, Brussel, pp. 327–61.

De Genst, W., 1994, *Ontwikkeling en implementatie van een systeem voor automatische afbeeldingsselektie*, Lic. Thesis, Geografisch Instituut, Vrije Universiteit Brussel, Brussel.

De Genst, W. and Canters, F., 1996, Development and implementation of a procedure for automated map projection selection, *Cartography and Geographic Information Systems*, **23**, 145–71.

Dent, B.D., 1975, Communicative aspects of value-by-area cartograms, *American Cartographer*, **2**, 154–68.

Dent, B.D., 1987, Continental shapes on world projections: the design of a poly centred oblique orthographic world projection, *Cartographic Journal*, **24**, 117–24.

Depuydt, F., 1983, The equivalent quintuple projection, *International Yearbook of Cartography*, **23**, 63–74.

Dorling, D., 1993, Map design for census mapping, *Cartographic Journal*, **30**, 167–83.

Dorling, D., 1995, *Area Cartograms: Their Use and Creation*, Department of Geography, University of Newcastle upon Tyne, Newcastle upon Tyne.

Dorling, D. and Fairbairn, D., 1997, *Mapping: Ways of Representing the World*, Harlow: Addison Wesley Longman Limited.

Dougenik, J.A., Chrisman, N.R. and Niemeyer, D.R. 1985, An algorithm to construct continuous area cartograms, *Professional Geographer*, **37**, 75–81.

Doytsher, Y. and Shmutter, B., 1981, Transformation of conformal projections for graphical purposes, *Canadian Surveyor*, **35**, 395–404.

Driencourt, L. and Laborde, J., 1932, *Traité des Projections des Cartes Géographiques à l'Usage des Cartographes et des Géodésiens*, 4 fascicules, Paris: Hermann et Cie.

Eckert, M., 1906, Neue Entwürfe für Erdkarten, *Petermanns Mitteilungen*, **52**, 97–109.

Eckert-Greifendorff, M., 1935, Eine neue flächentreue (azimutaloide) Erdkarte, *Petermanns Mitteilungen*, **81**, 190–2.

Erdi-Krausz, G., 1968, Combined equal-area projection for world maps, *Hungarian Cartographical Studies*, Budapest: Földmérési Intézet, pp. 44–9.

European Science Foundation (Ed.), 1992, *Geographic Information Systems: Data Integration and Data Base Design, A Proposal for an ESF Scientific Programme in the Social Sciences*, Jan. 1993 – Dec. 1996.

Fisher, I. and Miller, O.M., 1944, *World Maps and Globes*, New York: Essential Books.

Francula, N., 1971, *Die vorteilhaftesten Abbildungen in der Atlaskartographie*, Dipl. Ing. Dissertation, Institut für Kartographie und Topographie, Bonn.

Francula, N., 1980, Uber die Verzerrungen in den kartographischen Abbildungen, *Kartographische Nachrichten*, **30**, 214–16.

Freeman, H., 1991, Computer name placement, in Maguire, D.J., Goodchild, M.F. and Rhind, D.W. (Eds.), *Geographical Information Systems, Vol. 1: Principles*, Harlow: Longman, pp. 445–56.

Garver, J.B., 1988, New perspective on the world, *National Geographic*, **174**, 910–13.

Gilbert, E.N., 1974, Distortion in maps, *Society for Industrial and Applied Mathematics Review*, **16**, 47–62.

Gilmartin, P., 1985, The design of journalistic maps: purposes, parameters and prospects, *Cartographica*, **22**, 1–18.

Gilmartin, P.P., 1991, Showing the shortest routes – great circles, in American Cartographic Association (Ed.), *Matching the Map Projection to the Need*, Bethesda: American Congress on Surveying and Mapping, pp. 18–19.

Ginzburg, G.A., 1952, Kartograficheskiye proyektsii s izokolami v vide ovalov I ovoid, *Sbornik Statey po Kartografii*, **1**, 39–48.

Ginzburg, G.A. and Salmanova, T.D., 1962, Primeneniye v matematicheskoy kartografii metodov chislennogo analiza, TsNIIGAiK, *Trudy*, **153**, 6–80.

González-López, S., 1995, Conformal map projections by least squares adjustment with conditions between parameters, in *Proceedings of the 17th International Cartographic Conference*, Barcelona, pp. 776–80.

Goode, J.P., 1917, A new idea for a world map: a substitute for Mercator's projection, *Annals of the Association of American Geographers*, **7**, 75–6.

Goode, J.P., 1925, The Homolosine projection, a new device for portraying the Earth's surface entire, *Annals of the Association of American Geographers*, **15**, 119–25.

Goussinsky, B., 1951, On the classification of map projections, *Empire Survey Review*, **11**, 75–9.

Grafarend, E. and Niermann, A., 1984, Beste echte Zylinderabbildungen, *Kartographische Nachrichten*, **34**, 103–7.

Grave, D.A., 1896, *Ob osnovnykh zadachakh matematicheskoy teorii postroyeniya geograficheskikh kart*, St. Petersburg: Academy of Science.

Griffin, T.L.C., 1983, Recognition of areal units on topological cartograms, *American Cartographer*, **10**, 17–29.

Gusein-Zade, S.M. and Tikunov, V.S., 1993, A new technique for constructing continuous cartograms, *Cartography and Geographic Information Systems*, **20**, 167–73.

Györffy, J., 1990, Anmerkungen zur Frage der besten echten Zylinderabbildungen, *Kartographische Nachrichten*, **40**, 140–6.

Hägerstrand, T., 1957, Migration and area, *Lund Studies in Geography, Ser. B, Human Geography*, 13, Department of Geography, The Royal University of Lund, Sweden, pp. 27–158.

Haggett, P., 1965, *Locational Analysis in Human Geography*, London: Edward Arnold.

Hagiwara, Y., 1984, A retro-azimuthal projection of the whole sphere, in *International Cartographic Association, 12th Conference, Technical Papers*, Vol. 1, Perth, pp. 840–8.

Hammer, E., 1889a, *Über die geographisch wichtigsten Kartenprojektionen insbesondere die zenitalen Entwürfe, nebst Tafeln zur Verwandlung von geographischen Koordinaten in Azimutale*, Stuttgart: J.B. Metzlerschler.

Hammer, E., 1889b, Über Projektion der Karte von Afrika, *Zeitschrift der Gesellschaft für Erdkunde*, **24**, 222.

Hammer, E., 1892, Über die Planisphäre von Aitow und verwandte Entwürfe, insbesondere neue flächentreue ähnlicher Art, *Petermanns Mitteilungen*, **38**, 85–7.

Hammer, E., 1894, Die flächentreue Azimutal-Projektion für die Karte von Afrika, *Petermanns Mitteilungen*, **40**, 113–15.

Hammer, E., 1910, Gegenazimutale Projektionen, *Petermanns Mitteilungen*, **56–1**, 153–5.

Hatano, M., 1972, Consideration on the projection suitable for Asia-Pacific type world map and the construction of elliptical projection diagram, *Geographical Review of Japan*, **45**, 637–47.

Hoschek, J., 1969, *Mathematische Grundlagen der Kartographie*, Mannheim Zürich: Bibliographisches Institut.

Hsu, M.-L., 1981, The role of projections in modern map design, in L. Guelke (Ed.), *Maps in Modern Geography, Geographical Perspectives on the New Cartography*, Cartographica, Monograph 27, **18**, 151–86.

Hsu, M.-L. and Voxland, P.M., 1991, Showing routes for globe circlers, in American Cartographic Association (Ed.), *Matching the Map Projection to the Need*, Bethesda: American Congress on Surveying and Mapping, pp. 16–17.

Hunt, E.B. and Schott, C.A., 1854, Tables for projecting maps, with notes on map projections, *Report of the Superintendent of the Coast Survey 1853*, appendix no. 39, Washington: U.S. Coast Survey, pp. 96–163.

Jackson, J.E., 1968, On retro-azimuthal projections, *Survey Review*, **19**, 319–28.

James, H. and Clarke, A.R., 1862, On projections for maps applying to a very large extent of the Earth's surface, *London, Edinburgh and Dublin Philosophical Magazine*, 4th ser., **23**, 306–12.

Jankowski, P. and Nyerges, T.L., 1989, Design considerations for MaPKBS – map projection knowledge-based system, *American Cartographer*, **16**, 85–95.

Jiang, B. and Ormeling, F., 2000, Mapping cyberspace: visualizing, analysing and exploring virtual worlds, *Cartographic Journal*, **37**, 117–22.

Jordan, W., 1896, Der mittlere Verzerrungsfehler, *Zeitschrift für Vermessungswesen*, **25**, 249–52.

Kadmon, N., 1975, Data-bank derived hyperbolic scale equitemporal town maps, *International Yearbook of Cartography*, **15**, 47–54.

Kadmon, N. and Shlomi, E., 1978, A polyfocal projection for statistical surfaces, *Cartographic Journal*, **15**, 36–41.

Kaltsikis, C.J., 1989, Numerical treatment of conformal map projections, *Cartographic Journal*, **26**, 22–3.

Kavrayskiy, V.V., 1934, *Matematicheskaya Kartografiya*, Moscow–Leningrad.

Kavrayskiy, V.V., 1958, Izbrannyye trudy, vol. II: Matematicheskaja kartografija, part 1: Obscaja teorija kartograficheskih projekcij, *Izdaniye Nachal'nika Upraveleniya*, Gidrograficheskoy Sluzhby Voyenno-morskoy Akademii.

Keahey, T.A., 1999, Area-normalized thematic views. *Proceedings of the 19th International Cartographic Conference*, Ottawa, CD-ROM, section 06.

Kessler, F.C., 1991, *The Development and Implementation of MaPPS, An Expert System Designed to Assist in the Selection of a Suitable Map Projection*, Masters Thesis, Geography Department, Penn State University, State College, Pennsylvania.

Kidron, M. and Segal, R., 1984, *The New State of the World Atlas*, London: Pluto Press.

Kish, G., 1965, The cosmographic heart: cordiform maps of the 16th century, *Imago Mundi*, **19**, 13–21.

Kocmoud, C.J., 1997, *Constructing Continuous Cartograms: A Constraint-Based Approach*, Masters Thesis, Visualization Laboratory, Texas A&M University, College Station, Texas.

Laborde, J., 1928, *La nouvelle projection du Service Géographique de Madagascar*, Cahiers du Service Géographique de Madagascar, no. 1, Tananarive.

Laskowski, P., 1991, On a mixed local-global error measure for a minimum distortion projection, *ACSM Technical Papers*, American Congress on Surveying and Mapping, Vol. 2, pp. 181–6.

Laskowski, P., 1998, *The Distortion Spectrum*, Cartographica, Monograph 50, **34**.

Lee, L.P., 1944, The nomenclature and classification of map projections, *Empire Survey Review*, **8**, 142–52.

Lee, L.P., 1974, A conformal projection for the map of the Pacific, *New Zealand Geographer*, **30**, 75–7.

Lewis, C. and Campbell, J.D., 1951, *The American Oxford Atlas*, New York: Oxford University Press.

Littrow, J.J., 1833, *Chorographie; oder, Anleitung alle Arten von Land-, See- und Himmelskarten zu verfertigen*, Vienna: F. Beck.

McBryde, F.W., 1978, A new series of composite equal-area world map projections, *Abstracts of the 9th International Conference on Cartography*, College Park, Maryland, pp. 76–7.

McBryde, F.W. and Thomas, P.D., 1949, *Equal-area Projections for World Statistical Maps*, U.S. Coast and Geodetic Survey Special Publication, no. 245, Washington: U.S. Coast and Geodetic Survey.

Mackay, J.R., 1969, The perception of conformality of some map projections, *Geographical Review*, **59**, 373–87.

Maling, D.H., 1960, A review of some Russian map projections, *Empire Survey Review*, **15**, 203–15, 255–66, 294–303.

Maling, D.H., 1968, The terminology of map projections, *International Yearbook of Cartography*, **8**, 11–65.

Maling, D.H., 1992, *Coordinate Systems and Map Projections*, 2nd edition, Oxford: Pergamon Press.

Maurer, H., 1914, Die Definitionen in der Kartenentwürfslehre im Anschluss an die Begriffe zenital, azimutal und gegenazimutal, *Petermanns Mitteilungen*, **60**, 61–7, 116–21.

Maurer, H., 1919a, Das winkeltreue gegenazimutale Kartennetz nach Littrow (Weirs Azimutdiagramm), *Annalen der Hydrographie und maritimen Meteorologie*, **47**, 14–22.

Maurer, H., 1919b, "Doppelbüschelstrahlige, ortodromische" statt "doppelazimutale, gnomonische" Kartenentwürfe. Doppel-mittabstandstreue Kartogramme, *Annalen der Hydrographie und maritimen Meteorologie*, **47**, 75–8.

Maurer, H., 1935, *Ebene Kugelbilder, Ein Linnésches System der Kartenentwürfe*, Petermanns Mitteilungen, Ergänzungsheft no. 221.

Maurer, H., 1939, Kartennetze für meteorologische Zwecke; allgemeine Wetterkarten; neuartige breitenkreistreue (äquiparallele) Weltkarten, *Annalen der Hydrographie und maritimen Meteorologie*, **67**, 177–92.

Maurer, H., 1944, Erläuterung zur Weltkarte in breitenkreistreuem Entwürf nach Maurer, *Petermanns Geographische Mitteilungen*, **90**, 43–6.

Mekenkamp, P.G.M., 1990, The need for projection parameters in a GIS environment, *Proceedings of the First European Conference on Geographical Information Systems*, Amsterdam, pp. 762–9.

Miller, O.M., 1941, A conformal map projection for the Americas, *Geographical Review*, **31**, 100–4.

Miller, O.M., 1942, Notes on cylindrical world map projections, *Geographical Review*, **32**, 424–30.

Miller, O.M., 1953, A new conformal projection for Europe and Asia [sic; should read Africa], *Geographical Review*, **43**, 405–9.

Miller, O.M., 1955, *Specifications for a Projection System for Mapping Continuously Africa, Europe, Asia, and Australasia on a Scale of 1:5,000,000*, Contract Report to Army Map Service, May 1955, New York: American Geographical Society.

Miller, V.C., 1953, *A Quantitative Geomorphic Study of Drainage Basin Shape Characteristics in the Clinch Mountain Area, Virginia, and Tennessee*, Technical Report no. 3, Department of Geology, Columbia University, New York.

Monmonier, M., 1977, Nonlinear reprojection to reduce the congestion of symbols on thematic maps, *Canadian Cartographer*, **14**, 35–47.

Monmonier, M., 1989, *Maps With the News. The Development of American Journalistic Cartography*, Chicago: The University of Chicago Press.

Monmonier, M., 1990, *How to Lie with Maps*, Chicago: University of Chicago Press.

Monmonier, M., 1991, Centering a map on the point of interest, in American Cartographic Association (Ed.), *Matching the Map Projection to the Need*, Bethesda: American Congress on Surveying and Mapping, pp. 10–11.

Muehrcke, P.C., 1991, Showing ranges and rings of activity, in American Cartographic Association (Ed.), *Matching the Map Projection to the Need*, Bethesda: American Congress on Surveying and Mapping, pp. 24–5.

Muller, J.-C., 1978, The mapping of travel time in Edmonton (Alberta), *Canadian Cartographer*, **22**, 195–210.

Muller, J.-C., 1982, Non-Euclidean geographic spaces: mapping functional distances, *Geographical Analysis*, **14**, 189–203.

National Geographic Society, 1995, *National Geographic Atlas of the World*, Washington: National Geographic Society.

Nelder, J.A. and Mead, R., 1965, A simplex method for function minimization, *Computer Journal*, **7**, 308–13.

Nell, A.M., 1890, Aquivalente Kartenprojektionen, *Petermanns Mitteilungen*, **36**, 93–8.

Nestorov, I.G., 1997, CAMPREL: A new adaptive conformal cartographic projection, *Cartography and Geographic Information Systems*, **24**, 221–7.

Nyerges, T.L. and Jankowski, P., 1989, A knowledge base for map projection selection, *American Cartographer*, **16**, 29–38.

Olson, J.M., 1976, Noncontiguous area cartograms, *Professional Geographer*, **28**, 371–80.

Olson, J.M., 1991, Projecting the hemisphere, in American Cartographic Association (Ed.), *Matching the Map Projection to the Need*, Bethesda: American Congress on Surveying and Mapping, pp. 8–9.

Ortelius, A., 1570, *Theatrum orbis terrarum*, Facsimile edition, First series, vol. III, Amsterdam: Meridian Publishing Co, 1964.

Peters, A.B., 1975, Wie man unsere Weltkarten der Erde ähnlicher machen kann, *Kartographische Nachrichten*, **25**, 173–83.

Peters, A.B., 1978, Über Weltkartenverzerrungen und Weltkartenmittelpunkte, *Kartographische Nachrichten*, **28**, 106–13.

Poole, H., 1935, A map projection for the England–Australia air route, *Geographical Journal*, 86, 446–8.

Putnins, R.V., 1934, Jaunas projekcijas pasaules kartem, *Geografiski raksti, Folia Geographica III & IV*, Riga, 180–209.

Raisz, E., 1938, *General Cartography*, New York: McGraw-Hill.

Raisz, E., 1943, Orhoapsidal world maps, *Geographical Review*, **33**, 132–34.

Reeves, E.A., 1910, *Maps and Map-Making*, London: Royal Geographical Society.

Reilly, W.I., 1973, A conformal mapping projection with minimum scale error, *Survey Review*, **22**, 57–71.

Richardus, P. and Adler, R.K., 1972, *Map Projections for Geodesists, Cartographers and Geographers*, Amsterdam: North–Holland Publishing Company.

Robinson, A.H., 1951, The use of deformational data in evaluating world map projections, *Annals of the Association of American Geographers*, **41**, 58–74.

Robinson, A.H., 1974, A new map projection: its development and characteristics, *International Yearbook of Cartography*, **14**, 145–55.

Robinson, A.H., 1991, As the Earth turns, in American Cartographic Association (Ed.), *Matching the Map Projection to the Need*, Bethesda: American Congress on Surveying and Mapping, pp. 2–3.

Robinson, A.H., Sale, R.D., Morrison, J.L. and Muehrcke, P.C., 1984, *Elements of Cartography*, 5th edition, New York: John Wiley & Sons.

Selvin, S., Merrill, D., Sacks, S., Wong, L., Bedell, L. and Schulman, J., 1984, *Transformations of Maps to Investigate Clusters of Disease*, Lawrence Berkeley Laboratory, University of California, no. LBL–18550.

Shirrefs, W.S., 1992, Maps as communication graphics, *Cartographic Journal*, **29**, 35–42.

Siemon, K., 1935, Wegtreue Ortskurskarten, *Mitteilungen des Reichsamts für Landesaufnahme*, **11**, 88–95.

Siemon, K., 1936, Die Ermittlung von Kartenentwürfen mit vorgegebener Flächenverzerrung, *Deutsche Mathematik*, **1**, 464–74.

Siemon, K., 1937, Flächenproportionales Umgraden von Kartenentwürfen, *Mitteilungen des Reichsamts für Landesaufnahme*, **13**, 88–102.

Siemon, K., 1938, Flächenproportionales Umbeziffern der Punkte in Kartenentwürfen, *Mitteilungen des Reichsamts für Landesaufnahme*, **14**, 34–41.

Slocum, T.A., 1999, *Thematic Cartography and Visualization*, London: Prentice Hall.

Smith, D.G. and Snyder, J.P., 1989, Expert map projection selection system, *U.S. Geological Survey Yearbook, Fiscal Year 1988*, pp. 14–15.

Snyder, J.P., 1977, A comparison of pseudocylindrical map projections, *American Cartographer*, **4**, 59–81.

Snyder, J.P., 1978a, Equidistant conic map projections, *Annals of the Association of American Geographers*, **68**, 373–83.

Snyder, J.P., 1978b, The space oblique Mercator projection, *Photogrammetric Engineering and Remote Sensing*, **44**, 585–96.

Snyder, J.P., 1981a, Map projections for satellite tracking, *Photogrammetric Engineering and Remote Sensing*, **47**, 205–13.

Snyder, J.P., 1981b, The perspective map projection of the Earth, *American Cartographer*, **8**, 149–60 [see corrections in **9** (1982), p. 84].

Snyder, J.P., 1984a, A low-error conformal map projection for the 50 States, *American Cartographer*, **11**, 27–39 [see correction in Snyder, 1987a, p. 207].

Snyder, J.P., 1984b, Minimum-error map projections bounded by polygons, *Cartographic Journal*, **21**, 112–20 [see corrections in **22** (1985), p. 73].

Snyder, J.P., 1985, *Computer-Assisted Map Projection Research*, U.S. Geological Survey Bulletin, no. 1629, Washington: United States Government Printing Office.

Snyder, J.P., 1986, A new low-error map projection for Alaska, *ACSM Technical Papers, American Society for Photogrammetry and Remote Sensing, American Congress on Surveying and Mapping, Fall Convention*, Anchorage, 307–14.

Snyder, J.P., 1987a, *Map Projections: A Working Manual*, U.S. Geological Survey Professional Paper, no. 1395, Washington: United States Government Printing Office.

Snyder, J.P., 1987b, "Magnifying-glass" azimuthal map projections, *American Cartographer*, **14**, 61–8.

Snyder, J.P., 1988, New equal-area map projections for non-circular regions, *American Cartographer*, **15**, 341–55.

Snyder, J.P., 1993, *Flattening the Earth, Two Thousand Years of Map Projections*, Chicago: The University of Chicago Press.

Snyder, J.P., 1994, The Hill eucyclic projection, *Cartography and Geographic Information Systems*, **21**, 213–18.

Snyder, J.P. and Steward, H., 1988, *Bibliography of Map Projections*, U.S. Geological Survey Bulletin, no. 1856, Washington: United States Government Printing Office.

Snyder, J.P. and Voxland, P.M., 1988, *An Album of Map Projections*, U.S. Geological Survey Professional Paper, no. 1453, Washington: United States Government Printing Office.

Spilhaus, A.F., 1942, Maps of the whole world ocean, *Geographical Review*, **32**, 431–5.

Spilhaus, A. and Snyder, J.P., 1991, World maps with natural boundaries, *Cartography and Geographic Information Systems*, **18**, 246–54.

Sprinsky, W.H. and Snyder, J.P., 1986, The Miller oblated stereographic projection for Africa, Europe, Asia and Australasia, *American Cartographer*, **13**, 253–61.

Starostin, F.A., Vakhrameyeva, L.A. and Bugayevskiy, L.M., 1981, Obobshchennaya klassifikatsiya kartograficheskikh proyektsiy po vidu izobrazheniya meridianov i paralleley, *Izvestiya Vysshikh Uchebnykh Zavedeniy, Geodeziya i Aerofotos'emka*, **6**, 111–16.

Steers, J.A., 1970, *An Introduction to the Study of Map Projections*, 15th edition, London: University of London Press Ltd.

Stewart, J.Q., 1943, The use and abuse of map projections, *Geographical Review*, 33, 589–604.

Stirling, I.F., 1974, The new map projection, *New Zealand Cartographic Journal*, 4, 3–9.

Strebe, D.R., 1994, The generalized conic projection, *Cartography and Geographic Information Systems*, 21, 233–41.

Tissot, N.A., 1881, *Mémoire sur la Représentation des Surfaces et les Projections des Cartes Géographiques*, Paris: Gauthier Villars.

Tobler, W.R., 1962, A classification of map projections, *Annals of the Association of American Geographers*, 52, 167–75.

Tobler, W.R., 1963, Geographic area and map projections, *Geographical Review*, 53, 59–78.

Tobler, W.R., 1964, *Geographical Coordinate Computations, Part II, Finite Map Projection Distortions*, Technical Report no. 3, ONR Task no. 389–137, University of Michigan, Department of Geography, Ann Arbor.

Tobler, W.R., 1966a, Notes on two projections, *Cartographic Journal*, 3, 87–9.

Tobler, W.R., 1966b, Medieval distortions: The projections of ancient maps, *Annals of the Association of American Geographers*, 56, 351–61.

Tobler, W.R., 1973, A continuous transformation useful for districting, *Annals of the New York Academy of Sciences*, 219, 215–20.

Tobler, W.R., 1974, Local map projections, *American Cartographer*, 1, 51–62.

Tobler, W.R., 1977, Numerical approaches to map projections, in Kretschmer, I. (Ed.), *Beiträge zur theoretischen Kartographie*, Vienna: Franz Deuticke, pp. 51–64.

Tobler, W.R., 1986, Pseudo-cartograms, *American Cartographer*, 13, 43–50.

Torguson, J.S., 1990, *Cartogram: A Microcomputer Program for the Interactive Construction of Value-By-Area Cartograms*, Masters Thesis, Department of Geography, University of Georgia, Athens, Georgia.

Tsinger, N.Y., 1916, O naivygodneyskikh vidakh konicheskikh proyetsiy, *Izvestiya*, Akademiya Nauk, St. Petersburg, ser. 6, 10, 1693.

van der Grinten, A.J., 1904, Darstellung der ganzen Erdoberfläche auf einer kreisförmigen Projektionsebene, *Petermanns Mitteilungen*, 50, 155–9.

Wagner, K., 1932, Die unechten Zylinderprojektionen, *Aus dem Archiv der Deutschen Seewarte*, 51, 40.

Wagner, K., 1941, Neue ökumenische Netzentwürfe für die kartographische Praxis, *Jahrbuch der Kartographie*, pp. 176–202.

Wagner, K., 1944, Umformung von Mercator-Netzen, *Petermanns Geographische Mitteilungen*, 90, 299–306.

Wagner, K., 1949, *Kartographische Netzentwürfe*, Leipzig: Bibliographisches Institut.

Wagner, K., 1962, *Kartographische Netzentwürfe*, 2nd edition, Mannheim: Bibliographisches Institut.

Wagner, K., 1982, Bemerkungen zum Umbeziffern von Kartennetzen, *Kartographische Nachrichten*, 32, 211–18.

Watts, D.G., 1970, Some new map projections of the world, *Cartographic Journal*, 7, 41–6.

Weber, H., 1867, Ueber ein Princip der Abbildung der Theile einer krummen Oberfläche auf einer Ebene, *Journal für Reine und Angewandte Mathematik*, **67**, 229–47.

William-Olsson, W., 1968, A new equal area projection of the world, *Acta Geographica*, **20**, 389–93.

Winkel, O., 1921, Neue Gradnetzkombinationen, *Petermanns Mitteilungen*, **67**, 248–52.

Wray, T., 1974, The seven aspects of a general map projection, *Cartographica*, Monograph 11.

Wu, Z. and Yang, Q., 1981, A research on the transformation of map projections in computer-aided cartography, *Acta Geodetica et Cartographica Sinica*, **10**, 20–44.

Young, A.E., 1920, *Some Investigations in the Theory of Map Projections*, Royal Geographical Society, Technical Series, no.1, London: Royal Geographical Society.

Young, A.E., 1922, Two new map projections, *Geographical Journal*, **60**, 297–9.

Zöppritz, K.J. and Bludau, A., 1899, *Leitfaden der Kartenentwurfslehre für Studierende der Erdkunde und deren Lehrer*, 2nd edition, Leipzig.

Index of names

Subject index

Airy–Kavrayskiy criterion 43
Airy's least-squares criterion 42–3,
    48, 50, 137
Aitoff projection 24, 38, 53–4, 64–5,
    67, 94–5, 97, 101, 108, 128–9,
    205, 254, 281
Aitoff transformation 66, 128, 131–2
Aitoff–Wagner projection 53–4, 67,
    101, 108, 129–30, 182–7, 276,
    279, 281, 301, 303, 305, 307,
    309
American Cartographic Association
    2, 263
angular difference 90–1
animation 163, 167
area class maps 267
Armadillo projection 192
aspect of map projection 40, 50,
    71–82, 297
    change of 22, 71–3, 80–3, 115,
    252, 276, 292
    choice of 80, 221, 253, 260, 283,
    295
    diagonal 73, 75
    equioblique sub-aspects 75–6
    first transverse 73, 75–8, 80, 207
    impact on map projection
    properties 276
    normal (direct) 20, 22, 35,
    73–80, 82–3, 207, 209–10,
    212–13, 219
    oblique 73–5, 292
    optimisation of 50, 169, 176–7,
    207–12, 221–4, 230, 232, 234–7,
    249, 253, 295
    orthogonal 73

    plagal 73, 75–7, 207, 209–10,
    295
    second transverse 73, 75–7, 207
    simple oblique 73, 75–9, 207–12
    skew 73, 75–7, 207
    transverse 73–5
    transverse oblique 40, 73, 75–9,
    81, 207
Apianus' elliptical projection
    (Apianus II) 24, 38, 62–4, 121,
    254, 280
aphylactic projections 32
Arago's projection 24, 64, 67–8
Atlantis projection 40, 53, 55, 79–81
August's conformal projection 78
authalic projections 32
azimuthal projection
    equal-area 24, 38, 46–7, 92,
    146–8, 232, 256, 259, 275, 277,
    298–9
    equal-area (oblique) 147–8, 171,
    222, 224–48
    equal-area (transverse) 126,
    129–33
    equidistant 24, 38, 161, 255–6,
    259, 275, 277, 298–9
    equidistant (oblique) 222, 255,
    267–8, 270
    equidistant (transverse) 126–9
    gnomonic 24, 32, 38, 256, 259,
    269, 271, 275, 277, 298–9
    minimum-error 38, 170
    orthographic 24, 38, 85, 256,
    259, 269, 275, 277, 282, 298–9
    perspective 24, 170, 269, 275,
    277, 298–9

Printed in the United States
by Baker & Taylor Publisher Services